Randomization Tests

STATISTICS: Textbooks and Monographs

A SERIES EDITED BY

D. B. OWEN, Coordinating Editor

Department of Statistics
Southern Methodist University
Dallas, Texas

Randomization Tests

Second Edition

Eugene S. Edgington

Department of Psychology
University of Calgary
Calgary, Alberta, Canada

MARCEL DEKKER, INC.   New York and Basel

To Jane

Library of Congress Cataloging-in-Publication Data

Edgington, Eugene S., [date]
  Randomization tests.

  (Statistics, textbooks and monographs ; v. 77)
  Includes bibliographies and index.
  1. Statistical hypothesis testing. I. Title.
II. Series.
QA277.E32 1986   519.5'6   86-24044
ISBN 0-8247-7656-9

MARCEL DEKKER, INC.
270 Madison Avenue, New York, New York 10016

Current printing (last digit):
10   9   8   7   6   5   4   3   2   1

PRINTED IN THE UNITED STATES OF AMERICA

Preface to the Second Edition

This edition includes new material distributed throughout the book. The major differences between this and the first edition occur in the chapters numbered 1, 6, 11, and 12 in this edition.

Section 1.13, "Major Advances," was added to Chapter 1 for readers who wanted to learn something about the historical background of randomization tests. Reading the works of R. A. Fisher and E. J. G. Pitman in preparing Section 1.13 led to the realization that Fisher often has been given credit for contributions made by Pitman. For example, statisticians have described the "Fisher randomization test," applied to Darwin's data on plant length, as if it did not require the assumption of random sampling, and that is incorrect. Fisher's test required random sampling as well as random assignment; it was Pitman who pointed out the randomization tests could be performed in the absence of random sampling. One must turn to the works of Pitman, not Fisher, to find a theoretical basis for randomization or other permutation tests. Also outlined in Section 1.13 are major advances in theory and application that have been made since Fisher and Pitman.

In response to a widespread interest in randomization tests for factorial designs, Chapter 6 was completely rewritten and a computer program was added. The discussion of interactions was expanded to show how randomization tests sensitive to interactions may be conducted.

Chapter 11 is a new chapter that presents randomization tests that do not fit well into the other chapters. The tests are of practical value and also are useful for demonstrating the

iii

potential of the randomization test approach for increasing flexibility of experimental design and data analysis.

Chapter 12, on randomization test theory, is the most important addition to *Randomization Tests*. It provides a sound theoretical framework for understanding and developing randomization tests, based on a principle given in a 1958 article by J. H. Chung and D. A. S. Fraser.

Some of the randomization tests in this edition were developed with the financial support of the Natural Sciences and Engineering Research Council of Canada (Grant number A5571). Otto Haller wrote the two new computer programs, Programs 6.1 and 11.1.

Eugene S. Edgington

Preface to the First Edition

This book is intended primarily as a practical guide for experimenters on the use of randomization tests, although it can be used also as a textbook for courses in applied statistics. The book will be useful in all fields where statistical tests are commonly used. It is likely to be most relevant, however, in the fields of psychology, education, biology, and medicine. No special mathematical background is required to understand the book, but it is assumed that the reader has completed an introductory statistics course.

Experimental design books and others on the application of statistical tests to experimental data perpetuate the long-standing fiction of random sampling in experimental research. Statistical inferences are said to require random sampling and to concern population parameters. In experimentation, however, random sampling is very infrequent; consequently, statistical inferences about populations are usually irrelevant. Thus there is no logical connection between the random sampling model and its application to data from the typical experiment. The artificiality of the random sampling assumption has undoubtedly contributed to the skepticism of some experimenters regarding the value of statistical tests. What is a more important consequence of failure to recognize the prevalence of nonrandom sampling in experimentation, however, is overlooking the need for special statistical procedures that are appropriate for nonrandom samples. As a result, the development and application of randomization tests have suffered.

Randomization tests are statistical tests in which the data are repeatedly divided, a test statistic (e.g., t or F) is computed for each data division, and the proportion of the data divisions with as large a test statistic value as the value for the obtained results determines the significance of the results. For testing hypotheses about experimental treatment effects, random assignment but not random sampling is required. In the absence of random sampling the statistical inferences are restricted to the subjects actually used in the experiment, and generalization to other subjects must be justified by nonstatistical argument.

Random assignment is the only random element necessary for determining the significance of experimental results by the randomization test procedure; therefore assumptions regarding random sampling and those regarding normality, homogeneity of variance, and other characteristics of randomly sampled populations, are unnecessary. Thus, any statistical test, no matter how simple or complex, is transformed into a distribution-free test when significance is determined by the randomization test procedure. For any experiment with random assignment, the experimenter can guarantee the validity of any test he wants to use by determining significance by the randomization test procedure. Chapter 1 summarizes various advantages of the randomization test procedure, including its potential for developing statistical tests to meet the special requirements of a particular experiment, and its usefulness in providing for the valid use of statistical tests on experimental data from a single subject.

A great deal of computation is involved in performing a randomization test and, for that reason, such a means of determining significance was impractical until recent years, when computers became accessible to experimenters. As the use of computers is essential for the practical application of randomization tests, computer programs for randomization tests accompany discussions throughout the book. The programs will be useful for a number of practical applications of randomization tests, but their main purpose is to show how programs for randomization tests are written.

Inasmuch as the determination of significance by the randomization test procedure makes any of the hundreds (perhaps thousands) of published statistical tests into randomization tests, the discussion of application of randomization tests in this book cannot be exhaustive. Applications in the book have been selected to illustrate different facets of randomization tests so that the experimenter will have a good basis for generalizing to other applications.

   Randomization Tests is dedicated to my wife, Jane Armitage
Edgington, because of her encouragement and understanding.
It was written during sabbatical leave granted me by the University of Calgary. Many of my colleagues at the University have
been helpful, but three of them deserve a special thanks: Allan
Strain, for writing most of the programs; Otto Haller, for many
enjoyable discussions of randomization tests; and Gerard Ezinga,
to whom I am grateful for his careful reading of the entire manuscript. Finally, I want to express my indebtedness to Benjamin
S. Duran of Texas Tech University, whose critical and sympa-
thetic review was of inestimable value in revising the manuscript.

Eugene S. Edgington

Contents

List of Computer Programs

Randomization Tests

1
Introduction

1.1 RANDOMIZATION TESTS

A randomization test is a permutation test based on randomization (random assignment), where the test is carried out in the following manner. A test statistic is computed for the experimental data, then the data are permuted (divided or rearranged) repeatedly in a manner consistent with the random assignment procedure, and the test statistic is computed for each of the resulting data permutations. These data permutations, including the one representing the obtained results, constitute the reference set for determining significance. The proportion of data permutations in the reference set that have test statistic values greater than or equal to (or, for certain test statistics, less than or equal to) the value for the experimentally obtained results is the P-value (significance or probability value). If, for example, the proportion is 0.02, the P-value is 0.02, and the results are significant at the 0.05 but not the 0.01 level of significance. Determining significance on the basis of a distribution of test statistics generated by permuting the data is characteristic of all permutation tests; it is when the basis for permuting the data is random assignment that a permutation test is called a randomization test.

The preceding definition is broad enough to include procedures called randomization tests that depend on random sampling as well as randomization. The modern conception of a randomization test, however, is a permutation test that is based on randomization alone, where it does not matter how the sample is selected. It is this concept of randomization tests that will be the concern of this book.

The null hypothesis for a randomization test is that the obtained measurement for each experimental unit (i.e., person, family, plot of land, or other unit that is randomly assigned) will be the same under one assignment to treatments as under any alternative assignment that could have resulted from the random assignment procedure. Thus, when the null hypothesis H_0 is true, random assignment of experimental units to treatments randomly divides the measurements among the treatments. Each data permutation in the reference set represents the results that, if the null hypothesis is true, would have been obtained for a particular assignment.

1.2   NUMERICAL EXAMPLES

We will now consider an example to show how statistical inferences can be drawn for the subjects in a randomized experiment in the absence of random sampling. In the example, a t test will be used. Because the employment of the significance table for t is invalid in the absence of random sampling, significance will be determined by repeatedly computing t for all permutations of the data. In other words, a randomization test will be used to determine significance of t.

Example 1.1

An experimenter is interested in comparing the effectiveness of treatments A and B on reaction time. The requirements of the experimental task being rather complex, he is very careful in selecting the subjects, obtaining a total of 10 subjects. He randomly assigns five to each of the treatments and runs the experiment. An independent t test is conducted, using the formula

$$t = \frac{\overline{X}_A - \overline{X}_B}{(s^2/n_A + s^2/n_B)^{1/2}} \tag{1.1}$$

where

$$s^2 = \frac{\Sigma(X_A - \overline{X}_A)^2 + \Sigma(X_B - \overline{X}_B)^2}{n_A + n_B - 2} \tag{1.2}$$

The t value obtained for the data is 3.450. The experimenter is reluctant to determine the significance of t by using t tables because of his method of selecting subjects. Therefore, he decides to derive a theoretical distribution of t which does not involve the assumption of random sampling. To do this he divides the 10 reaction times in every possible way between treatments A and B with the restriction that each treatment must have five reaction times. There are 252 permutations of this kind. For each of the 252 data permutations, formula (1.1) is used to determine the value of t. Ten of the 252 permutations provide a t value as large as 3.450, so the P-value is 10/252, or about 0.04. Therefore the results are significant at the 0.05 level.

The logical justification for this procedure for determining the significance of the t value is straightforward. The null hypothesis H_0 is that the reaction time for every subject is independent of the treatment assignment. The random assignment of subjects to treatments allowed 252 equally probable ways in which the subjects could be assigned. If H_0 is true, a subject's reaction time would have been the same if he had been assigned to the alternative treatment. Thus, given the random assignment of subjects in conjunction with H_0, there are 252 equally probable ways in which the 10 reaction times could have been divided between the two treatments. If H_0 is true, how likely would it be that the random assignment performed in the experiment would provide one of the 10 largest values in the distribution of 252 values? The answer is 10/252, or about 0.04. Thus, the experimenter can conclude that such strong evidence against H_0 as is provided by his experimental results would occur no more than four times in 100 with a true H_0. Or in terms of conventional significance levels, he can say that his results are significant at the 0.05 level.

Significance tables for t give the values that a t statistic must reach in order to be significant at various conventional α levels, given that the assumptions (including random sampling) underlying the significance table are valid. The tabled values required for significance are called *critical values*. The standard t table gives a critical value of 2.306 for $\alpha = 0.05$, for a two-tailed test with 8 degrees of freedom. The critical value in the table for $\alpha = 0.01$ is 3.355. Thus use of the t table could lead to the belief that the obtained t value of 3.450 was significant at the 0.01 level inasmuch as the value is larger than the critical value for $\alpha = 0.01$, whereas the randomization test procedure gave a significance value of 0.04. In the absence of random sampling, use of the t table to determine significance is valid only to the extent that the significance given by the t table approximates that given by the randomization test procedure. Sometimes the approximation is good and sometimes it is poor; sometimes the discrepancy between the significance given by the two methods is in one direction and sometimes it is in the other. There are many factors that determine how closely the α level reached by experimental results according to a significance table approximates the significance level given by data permutation for those results.

Example 1.2

Let us now consider an example with few enough permutations to make it practical to list all of them. Five subjects are assigned randomly to two treatments, three subjects to treatment A and two to treatment B, and these results are obtained from the experiment: A: 18, 30, 54; B: 6, 12. For a two-tailed test, $|t|$, the absolute value of t, is used as the test statistic. Application of formula (1.1) to the data gives a t value of 1.81. To determine the significance of the obtained t value, t is computed by

TABLE 1   Data Permutations

	A	B	A	B	A	B	A	B	A	B
	6	30	6	18	6	18	6	12	6	12
	12	54	12	54	12	30	18	54	18	30
	18	—	30	—	54	—	30	—	54	—
$\overline{X} =$	12	42	16	36	24	24	18	33	26	21
t =	−3.00		−1.22		0.00		−0.83		0.25	
	A	B	A	B	A	B	A	B	A	B
	6	12	12	6	12	6	12	6	18	6
	30	18	18	54	18	30	30	18	30	12
	54	—	30	—	54	—	54	—	54	—
$\overline{X} =$	30	15	20	30	28	18	32	12	34	9
t =	0.83		−0.52		0.52		1.22		1.81	

formula (1.1) for all permutations of the data, providing the results shown in Table 1. Table 1 provides the following ordered theoretical sampling distribution of $|t|$: 3.00, 1.81, 1.22, 1.22, 0.83, 0.83, 0.52, 0.52, 0.25, 0. It can be seen that only two of the 10 values of $|t|$ are as large as 1.81; they are the values for the obtained experimental data and for one other data permutation. The two-tailed probability associated with the experimental results is the proportion of the data permutations with a value of $|t|$ as large as 1.81, and so the P-value is 0.20. The obtained value of $|t|$ is significant at the 0.20 level.

1.3   RANDOMIZATION TESTS AND NONRANDOM SAMPLES

A randomization test is valid for any kind of sample, regardless of how the sample is selected. This is an extremely important property because the use of nonrandom samples is common in experimentation, and parametric statistical tables (e.g., t and F tables) are not valid for such samples. [The discussion of non-random samples by Edgington (1966) is the principal source of the views in this section and in Sections 1.4 to 1.6.]

Parametric statistical tables are applicable only to random samples, and the invalidity of application to nonrandom samples is widely recognized. [For example, Hays (1972, p. 292) states: "All the techniques and theory that we will discuss apply to random samples, and do not necessarily hold for any data collected in any way."] The random sampling assumption underlying the significance tables is that of a sampling procedure that gives all possible samples of n individuals within a specified population the same probability of being drawn. The tables are based on theoretical distributions for all equally probable samples. Arguments regarding the "representativeness" of a nonrandomly selected sample are irrelevant to the question of its randomness: a random sample is random because of the sampling procedure used to select it, not because of the composition of the sample. Thus random selection is necessary to ensure that samples are random.

Violations of the random sampling assumption can take various forms. One violation is using all available subjects, without employing a sampling procedure of any kind. A second kind of violation is the systematic selection of certain kinds of subjects. A third form of violation can occur even when random sampling

has taken place: making statistical inferences about one population on the basis of random samples from another population,
such as a subpopulation. Therefore it is essential to the validity of parametric statistical tables to ensure that not only has a
population been sampled randomly but also that the sampled population is the one for which statistical inferences are drawn.

1.4   THE WIDESPREAD USE OF NONRANDOM
SAMPLES IN EXPERIMENTS

It must be stressed that violation of the random sampling assumption invalidates parametric statistical tables not just for the occasional experiment but for virtually all experiments. A person
conducting a poll may be able to enumerate the population to be
sampled and select a random sample by a lottery procedure, but
an experimenter has not enough time, money, or information to
take a random sample of the population of the world in order to
make statistical inferences about people in general. Few experiments in biology, education, medicine, psychology, or any other
field use randomly selected subjects, and those that do usually
concern populations so specific as to be of little interest. Whenever human subjects for psychological experiments are selected
randomly, often they are drawn from a population of students
who attend a certain university, are enrolled in a particular
class, and are willing to serve as subjects. Biologists and others
performing experiments on animals generally do not even pretend
to take random samples, although they use standard hypothesis-
testing procedures designed to test null hypotheses about populations. These well-known facts are mentioned here as a reminder of the rareness of random samples in experimentation and of
the specificity of the populations on those occasions when random
samples are taken. [For similar views on the widespread use of
nonrandom samples in experiments, see Cotton (1967, pp. 64–65),
Keppel (1973, p. 28), Kirk (1968, p. 13; 1978, pp. 259–260),
and Spence et al. (1976, pp. 173–175).]

1.5   THE IRRELEVANCE OF RANDOM SAMPLES
FOR THE TYPICAL EXPERIMENT

Of course, if experimenters were to start taking random samples
from the populations about which they want to make statistical
inferences, the random sampling assumption underlying significance

tables would be met, and provided that the relevant parametric assumptions (such as normality and homogeneity of variance) concerning the populations were met, parametric statistical tables could be employed validly for determining significance. The irrelevance of random samples for the typical experiment, however, makes that prospect unlikely.

In most experimentation the concept of population comes into the statistical analysis because it is conventional to discuss the results of a statistical test in terms of inferences about populations, not because the experimenter has sampled randomly some population to which he wishes to generalize. The population of interest to the experimenter is likely to be one that cannot be sampled randomly. Random sampling by a lottery procedure, a table of random numbers, or any other device requires a *finite population*, but experiments of a basic nature are not designed to find out something about a particular finite existing population. For example, with animals or human subjects the intention is to draw inferences applicable to individuals already dead and individuals not yet born, as well as those who are alive at present time. If we were concerned only with an existing population, we would have extremely transitory biological laws because every minute some individuals are born and some die, producing a continual change in the existing population. Thus the population of interest in most experimentation is not one about which statistical inferences can be made because it cannot be sampled randomly.

1.6   GENERALIZING FROM NONRANDOM SAMPLES

Statistical inferences about populations cannot be made without random samples from those populations, and random sampling of the populations of interest to the experimenter is likely to be impossible. Thus for practically all experiments valid statistical inferences about a *population* of interest cannot be drawn. In the absence of random sampling, *statistical* inferences about treatment effects must be restricted to the subjects (or other experimental units) used in an experiment. Inferences about treatment effects for other subjects must be *nonstatistical* inferences—inferences without a basis in probability. Nonstatistical generalization is a standard scientific procedure. We generalize from our experimental subjects to individuals who are quite similar in those characteristics that we consider relevant. For example, if the effects of a particular experimental treatment

depend mainly on physiological functions that are almost unaffected by the social or physical environment, we might generalize
from our experiments to persons from cultures other than the
culture of our subjects. On the other hand, if the experimental
effects were easily modified by social conditions, we would be
more cautious in generalizing to other cultures. In any event,
the main burden of generalizing from experiments always has
been, and must continue to be, carried by nonstatistical rather
than statistical logic.

1.7   DEFINITION OF RANDOMIZATION TEST

A statistical test for which the significance of experimental results is determined by permuting the data repeatedly to compute
t, F, or some other test statistic is called a *randomization test*.
If we consider a t test to be a particular way of computing a
test statistic, then a test can at one and the same time be a
t test and a randomization test if the significance of t is determined by the randomization test procedure. Randomization
tests, when conventional test statistics are computed, are not
alternatives to the conventional tests; rather, they *are* those
tests with the significance determined by a special procedure.
    Strictly speaking, the topic of this book is not a special
kind of statistical test but a method of determining statistical
significance that can be used in conjunction with all kinds of
tests. The randomization tests to be discussed illustrate the
nature and utility of the randomization test procedure of determining significance, which is the central topic of the book.

1.8   PRACTICALITY

In Example 1.2, there were only 10 permutations of the data;
consequently the amount of time required to determine significance by using a randomization test is not excessive. However,
the number of data permutations mounts rapidly with an increase
in sample size, so that even for fairly small samples the time
required with the use of a desk calculator may be prohibitive.
Electronic computers, however, make randomization tests practical.
In Chapter 3 it will be shown that when a computer installation
is accessible, the cost of a randomization test is negligible in
terms of time or money, even for large samples.

In order to provide experimenters with the background for making efficient use of computers for the performance of randomization tests, Chapter 3 gives suggestions for minimizing computer time and time to write programs. For many of the statistical tests used to illustrate the randomization test procedure of determining significance, suggested computer programs accompany the discussion to help the experimenter understand how to write his own programs or how to tell a programmer his or her requirements.

1.9   INTELLIGIBILITY

A P-value determined by a randomization test is easy to understand. Every bit of the computation to determine significance is done by the experimenter; the significance is obtained directly from calculating a proportion, instead of being read from a table. Another factor enhancing the intelligibility of randomization tests is that the correspondence between the method of permuting the data and the method of random assignment in the experiment provides the experimenter with a sound grasp of the interpretation to place on the P-values after they have been computed.

When significance tables are used to determine the significance of experimental results, the experimenter does not calculate the significance; rather it is determined from a table of critical values calculated by somebody else. By memorizing the assumptions associated with the use of the table for a particular statistical test, the experimenter may have a rough idea of what probability distribution the mathematician used to generate the table, but unless he or she has a strong mathematical background, it would not be possible to compute the distribution to generate the significance table. On the other hand, the probability distribution of the test statistic, whether it be t or F or some other statistic, is generated by the experimenter when a randomization test is used. The experimenter computes every test statistic value for the various data permutations and uses that theoretical distribution to compute the P-value associated with the experimental results. The "significance table" used with a randomization test (such as the distribution of 10 values of t in Table 1) is derived directly from the experimental data by computations performed by the experimenter. The experimenter has a complete understanding of the origin of the significance values.

The correspondence between the way in which data are permuted to determine significance and the way in which random assignment is carried out in an experiment permits an experimenter to see whether the method of permuting the data is appropriate for that experiment. We considered earlier the type of random assignment performed when an experimenter wants to use an independent t test, and we showed how the random assignment determined the data permutations involved in finding the significance. In experiments to be analyzed by repeated-measures analysis of variance and other techniques, other kinds of random assignment are involved, and correspondingly different ways of permuting the data are required. In every case, H_0 of no treatment effect is equivalent to the assumption that wherever a subject (or other assignment unit) is assigned, its measurement goes with it, and this H_0 is the justification for transforming possible random assignments into data permutations that could have been obtained.

The intelligibility of the randomization test procedure also is important to the people to whom the experimenter communicates the results of his analysis, many of whom may have little or no formal education in statistics. These may include the experimenter's employer, a government agency or other sponsoring organization, scientists interested in the research findings, and the general public. These persons may not know what a t test is, and although they can understand how significance is determined by the randomization test procedure, they would benefit from having an understanding of why one computes the strange test statistic called t for each data permutation. An advantage of randomization tests is that quite frequently a relatively complex test statistic reduces to a simple test statistic that is equivalent in the sense that it necessarily gives the same significance value for the randomization test procedure as the more complex test statistic. For both correlated and independent t tests, an equivalent test statistic to t is the difference between means, because the largest t is associated with the largest difference between means, the next smaller t with the next smaller difference between means, and so on. Thus, in explaining the results of the determination of the significance of t by the randomization test procedure to a person unfamiliar with t tests, it can be stated, without oversimplification, that the P-value is the proportion of data divisions giving a difference between group averages as large as the difference that was experimentally obtained. The existence of simple test statistics that are equivalent to more complex conventional test statistics is of considerable value in communicating the results of a randomization test to laypersons.

1.10   VALIDITY

Bradley (1968, p. 85) indicated that "eminent statisticians have
stated that the randomization test is the truly correct one and
that the corresponding parametric test is valid only to the extent
that it results in the same statistical decision." Sir Ronald
Fisher, who originated analysis of variance (ANOVA) techniques,
was one such eminent statistician, and Oscar Kempthorne is
another. Kempthorne (1955, p. 947) quoted Fisher (1936, p. 59)
as stating, in regard to determining significance by the randomization test procedure, that "conclusions have no justification
beyond the fact that they agree with those which could have
been arrived at by this elementary method." (Although this
article to which Kempthorne referred concerned nonexperimental
research, Fisher was referring to application of a procedure of
determining significance on the basis of permuting the data.)
At the beginning of the paragraph containing the Fisher quota-
tion, Kempthorne gave his own view of the validity of randomization tests: "Tests of significance in the randomized experiment have frequently been presented by way of normal law
theory, whereas their validity stems from randomization theory."

A review by Cotton (1973, p. 168) of Winer's (1971) book
on experimental design, while referring to the book as the best
of its kind, had a lengthy statement about the pervasiveness of
nonrandom samples and the need for reassessing the validity of
significance tables for such samples. Part of the statement
follows:

> Though Winer (p. 251) mentions randomization tests as a way
> of avoiding distribution assumptions, he is like most statis-
> tical textbook authors in neglecting a more important advan-
> tage: randomization tests permit us to drop the most implau-
> sible assumption of typical psychological research—random
> sampling from a specified population. We need to tell our
> students, again and again that random sampling occurs
> infrequently in behavioral research and that, therefore, any
> statistical tests making that assumption are questionable
> unless otherwise justified.

Cotton then pointed out again that randomization tests would
permit valid statistical inferences in the absence of random
sampling. He followed the lead of Fisher in emphasizing that
determining significance by the use of conventional published
statistical tables is of questionable validity until it has been

shown that the significance so obtained would agree closely with
that given by randomization tests. In the following passage,
Cotton (1973, p. 169) used the expression "randomization
theory" to refer to the theory of randomization tests:

> A rigorous approach by Winer would have been to indicate
> for which of his many designs the F tests are indeed
> justified by randomization theory and those for which an
> investigator should either make exact probability calculations,
> perform distribution-free tests which give approximately
> accurate significance levels, or perform F tests without pre-
> tending to meet their assumptions.

Cotton's remarks are relevant to many fields of research
because, in most fields where data are analyzed by the use of
statistical tests, the assumption of random sampling usually is
not met. That is the case in cloud-seeding experiments. A
Statistical Task Force (Brillinger, Jones and Tukey, 1978) was
established to advise the United States Government on the statistical analysis of weather modification data. In its endorsement
of the use of randomization tests, the Task Force made the
following statement:

> The device of judging the strength of evidence offered by
> an apparent result against the background of the distribu-
> tion of such results obtained by replacing the actual random-
> ization by randomizations that might have happened seems
> to us definitely more secure than its presumed competitors,
> that depend upon specific assumptions about distribution
> shapes or about independence of the weather at one time
> from that at another. (Page D-1)

In a later passage, the Task Force again stressed the unrealistic
nature of parametric assumptions in regard to weather modification and again pointed to the necessity of doing "re-randomiza-
tions" (i.e., performing randomization tests):

> Essentially every analysis not based on *re*-randomization
> assumes that the days (or storms) behave like random sam-
> ples from some distribution. Were this true in practice,
> weather could not show either the phenomenon of persistance
> (autocorrelation) or the systematic changes from month to
> month within a season which we know do occur. In an era
> when re-randomization is easily accessible, and gives

trustworthy assessments of significance, it is hard indeed
to justify using any analysis making such assumptions.
(Page F-5.)

If the use of significance tables always provided close
approximations to the significance values given by randomization
tests, the fact that randomization tests are the only completely
valid procedures to use for finding significance with nonrandom
samples would be of academic interest only. The significance
tables, however, sometimes provide values that differ considerably from those given by randomization tests, and so the consideration of validity is a practical issue. The P-values given
by the two methods, are, of course, sometimes very similar. In
cases of large differences, the difference is sometimes in one
direction and sometimes in the other. Too many factors affect
the closeness of the significance provided by the two methods
to allow formulation of a general rule for the experimenter to
tell him how much he would be in error by using significance
tables instead of randomization tests. Some of the factors are:
the absolute sample sizes, the relative sample sizes for the different treatments, the sample distribution shape, the number of
tied measurements, and the type of statistical test.

The preceding assertion of the invalidity of the standard
procedures for determining significance with *nonrandom* samples
should not be confused with the common assertions that violations of parametric assumptions render conventional tests, like
the t test and analysis of variance, invalid. The assumptions
for which the effects of violations have been studied are those
that concern *population* characteristics, like normality and homogeneity of variance. The questions those studies tried to answer
were of the following kind: "How much would the use of our
standard significance tables mislead us if the populations that
were randomly sampled had quite different variances instead of
the identical variances that are assumed by the tables?" Such
studies of "robustness" would be more relevant if, like studies
by Collier and Baker (1966) and by Hack (1958), they dealt with
characteristics of samples instead of characteristics of populations.
When an experimenter deliberately selects subjects for an experiment there is no defined population, and so assumptions about
normality or other characteristics of the population are meaning-
less.

Despite the acknowledged validity of randomization tests, it is
not anticipated that they will supplant significance tables as the
most common means of determining significance. The reason is

that, for routine use of commonly used tests like the t test and one-way ANOVA, the validity of the significance tables is seldom challenged. In the less common applications of statistical tests, however, questions of validity arise, and it is for such situations that determination of significance by a randomization test is especially valuable. The application of ANOVA to dichotomous data, for example, is likely to arouse concern over validity in situations where application to quantitative data would not do so. The validity issue also comes to the fore if an experimenter wants to apply ANOVA to data from a single-subject experiment. With less commonly used tests, like analysis of covariance and multivariate analysis of variance, any kind of application may stimulate discussions of the tenability of the assumptions underlying the statistical tables. Certainly when first a new statistical test is introduced, its validity must be demonstrated. As a consequence of the multiplicity of ways in which the validity of P-values can be called into question, it is not unusual for an experimenter to use a less powerful statistical test than the one he has in mind, or even to avoid experiments of a certain kind (such as single-subject experiments), in order to have publishable results. Compromises of this kind can be avoided by using randomization tests, instead of conventional statistical tables, for determining significance.

1.11   VALIDITY AND EXPERIMENTAL INDEPENDENCE

In addition to randomization, experimental independence is a major requirement for ensuring the validity of randomization tests of treatment effects. Two subjects are experimentally independent if one subject does not influence the measurement of the other subject. The following example will show how P-values can be spuriously small when there is lack of experimental independence.

Example 1.3

Ten rats are assigned randomly to treatments with five rats being assigned to a treatment group that is vaccinated against a certain disease and the other five being assigned to a control group which is not vaccinated. Unknown to the experimenter, one of the rats already had the disease when it was assigned and the disease spread throughout the group to which it was assigned. That rat was assigned to the control group, and even though the rats in that group were kept in separate cages, the disease was transmitted by a laboratory assistant who sometimes handled the rats. Thus, all of the control rats caught the disease. The disease was not transmitted to any of the vaccinated rats because they were housed in a separate room, served by a different laboratory assistant. A randomization test based on the prediction of a higher rate of disease for the control group would give a P-value of 1/252, because there are 10!/5!5! = 252 ways to divide five diseased and five healthy rats between the control and vaccinated groups, and in only one of those divisions are all five diseased rats in the control group and all five healthy rats in the experimental group. To see how misleading the P-value can be, consider that the chance of assigning the diseased rat to the control group and having the disease spread to the other control animals was 1/2, and that assignment alone would have resulted in the disease in all of those animals, providing a P-value of 1/252, even if the vaccination had no effect whatsoever. With absolutely identical treatments, carryover effects of this kind could result in a high probability of getting a small P-value.

Aside from contagious diseases, there are many other ways in which experimental independence can be violated by making the responses of an experimental subject dependent on the other subjects assigned to that treatment condition. For instance, in an experiment on psychological effects of semistarvation, if the food-deprived subjects are not kept isolated from each other, the behavior of one or two subjects in the group may greatly influence the behavior of the others, violating the assumption of experimental independence and invalidating statistical tests.

It is sometimes difficult to isolate subjects from each other sufficiently to ensure experimental independence. For example, in a comparison of different teaching methods, it is likely that students within classrooms influence the learning of other students in the same classrooms. Hundreds of students may be assigned at random to the classrooms taught by the different methods, but the results as a whole may be strongly dependent on just a few students. If one or two students in a classroom, by their enthusiasm, boredom, penetrating questions, or other means, strongly influence the learning of the rest of the students in their class, the lack of experimental independence could make the results almost worthless.

Experimenters should isolate experimental subjects from each other, or by some other means, reduce chances of communication or other interaction that may result in one subject's measurements being influenced by other experimental subjects. Although it may be impossible to completely rule out interaction among subjects, minimizing the opportunity for interaction should be a consideration in designing an experiment.

1.12   VERSATILITY

Randomization tests are extremely versatile as a consequence of their potential for ensuring the validity of existing statistical tests and for developing new special-purpose tests. Intelligibility to nonstatisticians is a property of randomization tests that enables the experimenter to actualize their potential and thereby acquire greater flexibility in planning his experiment and analyzing the experimental data.

Determining significance by permuting the data is a distribution-free procedure, and so any statistical test for which the significance is determined in this way is a distribution-free test. Parametric tests of all kinds, including relatively complex tests like factorial analysis of variance, analysis of covariance, and multivariate analysis of variance, are distribution-free tests when the significance of the test statistic is determined by permuting the data. Thus, given random assignment, an experimenter guarantees the validity of a statistical test when he uses a randomization test procedure to determine significance.

The idea of a researcher possessing both the expertise and the time to develop a new statistical test to meet his special research requirements may sound farfetched, as indeed it would be if a special significance table had to be constructed for the new test. The original derivation of the sampling distribution of a new test statistic preparatory to formulation of a significance table is time consuming and demands a high level of mathematical ability. On the other hand, if the researcher determines significance by data permutation, not much time or mathematical ability is necessary, because the "significance table" is generated by data permutation. Examples of new statistical tests with significance based on the randomization test procedure (data permutation) are dispersed throughout the following chapters.

The versatility of randomization tests is related to increased power or sensitivity of statistical testing in a number of ways.

That is, there are several ways in which the versatility may increase the chances of an experimenter getting significant results when there are actual differences in treatment effects. For example, ANOVA procedures and others that are more powerful than rank order tests are sometimes passed over in favor of a rank order test in order to ensure validity, whereas validity can be guaranteed by using randomization tests without the loss of power that comes from transforming measurements into ranks. In cases where there is no available rank order counterpart to a parametric test whose validity is in question, a randomization test can be applied to relieve the test of parametric assumptions, including the assumption of random sampling. The possibility of ensuring the validity of any existing statistical test, then, provides the experimenter with the opportunity to select the test that is most likely to be sensitive to the type of treatment effect he expects. Furthermore, he is not limited to existing statistical tests; use of randomization tests allows the experimenter to develop statistical tests that may be more sensitive to treatment effects than conventional ones. The experimental design possibilities also are greatly expanded because new statistical tests can be developed to accommodate radically different random assignment procedures.

In light of the multitude of uses for randomization tests, little more can be done in this book than to present a few of the possible applications to illustrate the versatility of this special means of determining significance. Those applications are to statistical tests of various kinds, and experimenters should encounter no difficulty in generalizing to other statistical tests.

1.13 MAJOR ADVANCES

André (1883) developed a recursion formula which is the basis for a runs test, generating a distribution of number of runs (sequences of ascending or descending values) for N! orders of a sequence of N numerical values. The resulting probability distribution can be used to perform a permutation test on the basis of either random sampling or random assignment. As the number of runs depends only on the ranks of the numerical values and not on the actual values, it has been possible to provide a table for this permutation test, which is given in Bradley (1968, p. 363).

Fisher (1935, Section 21) carried out a permutation test which employed a reference set of test statistic values dependent on

the actual values of the measurements involved, not just their
ranks. Apparently, this was the first permutation test employing a reference set that had to be derived separately for each
different set of data; unlike earlier permutation tests which
utilized only the ordinal properties of the data, this test did
not allow the derivation of general purpose significance tables.
The hypothetical experiment and test were described by Fisher
as follows. Fifteen seeds were selected randomly from each of
two populations to test the null hypothesis of identity of the
populations with respect to the size of plants the seeds would
produce under the same conditions. The seeds could not be
grown under identical conditions, and so, in order to prevent
population differences from being confounded with soil and
environmental differences, the seeds in the samples were assigned randomly to the pots in which they were to be planted.
Pots with similar soil and locations were paired, and for each of
the 15 pairs of pots it was randomly determined which one of a
pair of seeds would be planted in each pot. After the plants
reached a certain age, the heights were measured and the
measurements were paired as they would be for a correlated or
paired t test. Fisher, however, carried out a permutation test.
He considered the reference set of $2^{15} = 32,768$ data permutations associated with switching the signs of the differences for
all 15 pairs and computed the difference between totals for the
two types of seeds as the test statistic. The P-value was the
proportion of those data permutations with as large a difference
between totals as the obtained difference. It is a credit to
Fisher's analytical ability that he could determine what proportion of the data permutations provided as large a difference
between totals as the obtained value without computing the test
statistic for all 32,768 data permutations. That it was a difficult
task, even for Fisher, is reflected in his need to revise the
randomization test P-value in later editions of *Design of Experiments*.

Fisher's test frequently has been discussed as if it, like
modern randomization tests, was based on random assignment
alone. For example, a series of papers by six authors on "the
Fisher randomization test" headed by a paper by Basu (1980)
discussed the above test as if no random sampling was involved,
which is inconsistent with Fisher's description of the test.
Fisher at several points in his discussion referred to testing
the null hypothesis of identity of populations and, in fact, random sampling of populations was required for his test. Without
random selection of the seeds, the null hypothesis tested would

have been identity of the effect of the paired pots on plant growth, an effect of no interest to Fisher. The random assignment in Fisher's example was to control for, not test, differences between pots. Fisher's test was valid for his purpose, however, which was to compare two randomly sampled populations, not simply random assignment of a finite number of seeds to soil conditions. The experimental units (seeds) were not assigned at random for the purpose of determining the effect of the pots they were assigned to, and the switching of signs of differences between measurements within pairs was not to represent what would have resulted under the null hypothesis if the assignment of seeds within a pair had been reversed. It was the null hypothesis of identical infinite populations that justified changing the sign of the difference between paired observations: random sampling of the populations plus random assignment within pairs ensured that any absolute difference between totals had a plus or a minus sign with equal probability. We see then that the Fisher randomization test, unlike modern randomization tests, did not involve random assignment to test the effect of experimental treatments and did require random sampling of infinite populations.

Pitman (1937, 1938) provided the first theoretical framework for randomization tests and nonexperimental permutation tests. In his first paper on permutation tests, Pitman (1937a, p. 119) disclaimed priority of his theory by this cautious statement: "The main idea is not new, it *seems* to be *implicit* in all Fisher's writings" (italics added). Pitman (1937a, b, 1938) gave several examples of permutation tests as well as a rationale applicable both to randomization tests and nonexperimental permutation tests. After providing a rationale for permutation tests of differences between populations, he showed that permutation tests could be based on random assignment alone, and this demonstration was of great importance to the large number of experimenters who use random assignment but not random sampling (Pitman, 1937a, p. 129).

A large number of rank tests were developed shortly after the articles by Pitman. Rank tests tended to be represented as testing hypotheses about populations on the basis of random samples, even in regard to experimental research. These rank permutation tests have been used widely because of their readily available significance tables. The applicability of rank tests to experimental data on the basis of random assignment alone tended to be overlooked, as was the possibility of permuting raw data in the same way as the ranks are permuted in order to have a more sensitive test. Lehmann (1975), in his book on non-parametric tests, is exceptional in correcting these oversights.

While permutation tests for ranks were being developed within the random sampling framework, there was work being done in regard to permutation tests based on randomization, that is, randomization tests. There was no flurry of interest in randomization tests to parallel the rapid development of the rank test field, however. The primary use of randomization tests was to stimulate re-examination of analysis of variance and other normal curve procedures for analyzing experimental data. Kempthorne (1952, 1955) and his colleagues, for instance, did considerable work on "randomization analysis," which concerns investigation of the null and non-null distributions of parametric test statistics (especially F) under experimental randomization. Randomization analysis reflected the attitude taken toward randomization tests by many early investigators: randomization tests were useful tools for assessing the validity of parametric tests under violation of the parametric test's assumptions. Liberation of randomization tests from that subordinate role, and recognition of them as valuable tests in their own right, was delayed until the accessibility of computers made them practical. Nevertheless, Kempthorne made an important contribution to the field of randomization tests by keeping the concept of the test in the public eye for a number of years. In addition, he provided an alternative to the "permutation" rationale for randomization tests. Kempthorne described the generation of what we have called data permutations as being accomplished by superimposing the set of obtained data upon each of the possible assignment patterns. For example, instead of using a data manipulation procedure to generate data divisions for a latin square design, the square array of data would be placed over each of the possible latin squares and the data in each case would be associated with the treatment designations in the latin square upon which the data were placed. This conceptualization, termed the *randomization-referral* approach (Chap. 12), is extremely useful for understanding the rationale of randomization tests.

Randomization tests came into their own with the advent of the computer age. Electronic computers, even microcomputers, can very quickly and inexpensively generate thousands of data permutations for performing a randomization test. Nevertheless, with only moderate sample sizes there may be so many possible assignments that it would not be feasible to provide data permutations for all possible assignments, even with modern computers. (There are, for example, over 5 trillion ways to assign 30 subjects to three treatments with 10 subjects per treatment.) Dwass (1957) helped solve this problem by providing a rationale for "modified randomization tests" based on random samples of all data permutations. The randomly generated reference set, for example, might consist of only 1000 of the data permutations. The randomization test procedure proposed by Dwass is not simply a Monte Carlo approximation to a "true randomization test," but is a completely valid test, even when the number of data permutations in the random subset is a very small portion of the number associated with all possible assignments. The use of this random data permuting procedure, which is valid for nonexperimental permutation tests as well as randomization tests, was facilitated by the construction of computer algorithms for random permuting of data by Green (1963, pp. 172–173) and others.

Another procedure designed to make randomization tests more practical by reducing the number of data permutations in the reference set was proposed by Chung and Fraser (1958). Their procedure also involved the use of only a subset of the permutations associated with all possible assignments, but, unlike Dwass's subset, theirs was a systematic (nonrandom) subset. It is doubtful that the Chung and Fraser approach, based on a permutation group rationale, will be as useful as the random procedure for reducing the computational time for a randomization test, but their permutation group concept is of considerable theoretical importance and is fundamental to the general randomization test theory presented in Chapter 12. Their approach provides a sound and general theoretical justification for many applications of randomization tests to complex experimental designs, thus providing a basis for a substantial expansion of the scope of applicability of randomization tests.

Only major contributions have been considered here, but it is sufficient to indicate that much progress has been made since the pioneering work of Fisher and Pitman in the 1930s. As people become more aware of the potential of randomization tests, contributions to the field will become more frequent.

REFERENCES

André, D. (1883). Sur le nombre de permutations de *n* éléments qui présentent *s* séquences. *Comptes Rendus (Paris)*, *97*, 1356–1358.

Basu, D. (1980). Randomization analysis of experimental data: the Fisher randomization test. *J. Am. Statist. Assn.*, *75*, 575–582.

Bradley, J. V. (1968). *Distribution-free Statistical Tests*. Prentice-Hall, Englewood Cliffs, NJ.

Brillinger, D. R., Jones, L. V., and Tukey, J. W. (1978). *The Management of Weather Resources. Volume II. The Role of Statistics in Weather Resources Management*. Report of the Statistical Task Force to the Weather Modification Board. Department of Commerce: Washington, DC.

Chung, J. H., and Fraser, D. A. S. (1958). Randomization tests for a multivariate two-sample problem. *J. Am. Statist. Assn.*, *53*, 729–735.

Collier, R. O., and Baker, F. B. (1966). Some Monte Carlo results on the power of the F-test under permutation in the simple randomized block design. *Biometrika*, *53*, 199–203.

Cotton, J. W. (1967). *Elementary Statistical Theory for Behavior Scientists*. Addison-Wesley, Reading, MA.

Cotton, J. W. (1973). Even better than before. *Contemp. Psychol.*, *18*, 168–169.

Dwass, M. (1957). Modified randomization tests for nonparametric hypotheses. *Ann. Math. Statist.*, *28*, 181–187.

Edgington, E. S. (1966). Statistical inference and nonrandom samples. *Psychol. Bull.*, *66*, 485–487.

Fisher, R. A. (1935). *Design of Experiments*. Oliver and Boyd, Edinburgh.

Fisher, R. A. (1936). The coefficient of racial likeness and the future of craniometry. *J. R. Anthropol. Inst.*, *66*, 57–63.

Green, B. F. (1963). *Digital Computers in Research*. McGraw-Hill, New York.

Hack, H. R. B. (1958). An empirical investigation into the distribution of the F-ratio in samples from two non-normal populations. *Biometrika*, *45*, 260–265.

Hays, W. L. (1972). *Statistics for the Social Sciences* (2d ed.). Holt, Rinehart and Winston, New York.

Kempthorne, O. (1952). *Design and Analysis of Experiments*. Wiley, New York.

Kempthorne, O. (1955). The randomization theory of experimental inference. *J. Am. Statist. Assn.*, *50*, 946–967.

Keppel, G. (1973). *Design and Analysis: A Researcher's Handbook*. Prentice-Hall, Englewood Cliffs, NJ.

Kirk, R. E. (1968). *Introductory Statistics*. Brooks/Cole, Belmont, CA.

Lehmann, E. L. (1975). *Nonparametrics: Statistical Methods Based on Ranks*. Holden-Day, San Francisco.

Pitman, E. J. G. (1937a). Significance tests which may be applied to samples from any populations. *J. R. Statist. Soc. B.*, *4*, 119–130.

Pitman, E. J. G. (1937b). Significance tests which may be applied to samples from any populations. II. The correlation coefficient. *J. R. Statist. Soc. B.*, *4*, 225–232.

Pitman, E. J. G. (1938). Significance tests which may be applied to samples from any populations. III. The analysis of variance test. *Biometrika*, *29*, 322–335.

Spence, J. T., Cotton, J. W., Underwood, B. J., and Duncan, C. P. (1976). *Elementary Statistics* (3d ed.). Appleton-Century-Crofts, New York.

Winer, B. J. (1971). *Statistical Principles in Experimental Design* (2d ed.). McGraw-Hill, New York.

2
Random Assignment

Simple applications of randomization tests can be made on the basis of a general understanding of the role of random assignment. To take full advantage of the versatility of randomization tests, however, the experimenter requires a more detailed discussion of random assignment than is provided in standard statistics books. The focus of this chapter is the part random assignment plays in guaranteeing the *validity* of randomization tests. How variations in random assignment procedures affect the *power* (sensitivity to treatment effects) of randomization tests will be dealt with in later chapters, for various types of statistical tests.

2.1 BETWEEN-SUBJECT AND WITHIN-SUBJECT VARIABILITY

The principal function served by random assignment is to allow one to take into account statistically the effects on experimental measurements resulting from subject differences and variation within subjects over time, so that statistical inferences can be drawn regarding the effects of the treatment per se.

Living organisms, the typical subjects in psychology, biology, medicine, and other social and biological sciences, are heterogeneous in many respects likely to be relevant to experimental outcomes. Subjects can be selected to ensure homogeneity for a few characteristics, but it would be impractical to select subjects homogeneous with regard to all relevant characteristics, even if they were known and readily ascertainable. Care must be taken, then, to make sure the differences between measurements under alternative treatments are not a function of between-subject variability that still remains after making the sample of subjects as homogeneous as possible in certain important respects. Between-subject variability must be taken into account in order to make valid statistical inferences about treatment effects. This may be done statistically or nonstatistically.

Apparently, the only nonstatistical way to ensure that differences between measurements from different treatments are not due to subject differences is to have each subject take every treatment, as in a standard repeated-measures design. With no differences whatsoever between the subjects taking the different treatments, between-subject variability is completely controlled.

A repeated-measures design, however, is not always possible. For example, in a study of visual development an animal could not be raised in total darkness for the first few months of life and also be raised with the usual amount of light during the same time period. Because it is frequently impossible (or, if possible, impractical) to have all subjects take all treatments, independent-groups experiments, where different subjects take different treatments, are important. There is no way of equating groups with respect to subject variables to provide complete control for individual differences in independent-groups experiments, because of the impossibility of ensuring complete homogeneity with respect to all relevant characteristics. Because there is no direct complete control over between-subject variability for independent-groups experiments, it is necessary to provide that control by random assignment of subjects to treatments.

In addition to between-subject variability, within-subject variability over time also must be controlled when living organisms are used as subjects in experiments. Individual subjects change in ways important to biological and social scientists even over short periods of time, regardless of the treatments to which they are assigned, and so care must be taken to ensure that such changes are not mistaken for treatment effects.

In independent-groups experiments, it is important not to have one treatment given at systematically better times than another. If one treatment is given at a time of day when subjects generally are more responsive, higher measurements for that treatment may simply reflect variability in responsiveness within subjects over the day rather than treatment differences. Ordinarily a number of different treatment times will be used in

an experiment. Systematic differences between treatment times
for different treatments can be avoided by assigning a treatment
time to a subject before assignment to a treatment. Random
assignment of a subject, along with a predetermined treatment
time, to a treatment provides statistical control over both
between-subject and within-subject variability.

For repeated-measures experiments where all subjects take
all treatments, each subject should have as many treatment times
as treatments allocated to him. Following that, there is random
assignment of a subject's treatment times to treatments. Having
each subject take all treatments controls for between-subject
variability, and random assignment of treatment times controls
for within-subject variability.

We have seen that random assignment in an experiment is
not restricted to random assignment of subjects; also possible
is the random assignment of treatment times to treatments. Furthermore, there can be random assignment of subjects-plus-treat-
ment-times, where each subject is given a treatment time, and
the subject along with that treatment time is assigned to a treatment. Subjects, treatment times, and subjects-plus-treatment-
times, then, constitute three kinds of entities that can be assigned randomly. The following section will treat the topic of experimental units in a more general way in order to provide the background for consideration of more complex experimental units.

2.2   EXPERIMENTAL UNITS

The term *sampling unit* is used in statistics to refer to units
that are selected from a population with random sampling. For
example, if there is random selection of cities from a population
of cities in a country, the sampling unit is a city, or if houses
are selected randomly from a population of houses in a city, the
sampling unit is a house. A sampling unit may be an individual
element, like a person or a tree, or it may be a group of elements, like a family or a forest.

The term *experimental unit* will be used to refer to analogous
units associated with random assignment. An experimental unit
is a unit that is assigned randomly to a treatment. In many
experiments the unit that is assigned is a subject (e.g., a person) or a group of subjects (e.g., a family), in conjunction with
a treatment time. In using a lottery for random assignment, that
which is designated by an individual pellet or slip of paper is
the experimental unit. When a slip of paper is drawn, the

corresponding unit, whether it consists of one or many subjects, is assigned in its entirety to the treatment.

In any experiment an experimental unit is assigned to only one treatment administration. Repeated-measures experiments are apparent exceptions to this rule because a single subject is assigned to several treatments in such experiments. But these experiments are not in fact exceptional, because a single subject is not a single experimental unit in a repeated-measures experiment; treatment times are assigned randomly to treatments for each subject, so a single experimental unit is a subject plus a treatment time.

2.3 COMPLEX EXPERIMENTAL UNITS

Where there is need to provide statistical control over both between- and within-subject variability, the basic experimental unit is subject-plus-treatment-time. More complex experimental units are sometimes used, and two types of them will be considered here.

One type of complex experimental unit differs from a simple unit in that a number of subjects, not just one, are included within a single experimental unit. For example, for practical reasons an educational psychologist may randomly assign intact classes, instead of individual students, to instructional procedures. If there were five classes, two of which were to be assigned to one procedure and three to another, there would be five experimental units, consisting of the five classes along with predetermined treatment times, perhaps extending over several weeks. No matter how many students were in each class, the total number of possible assignments would be only $5!/2!3! = 10$, the number of ways of dividing five things into two groups with two in a specified group and three in the other. Thus, although a large number of students could be involved, the smallest possible significance value that could be obtained is $1/10$.

Determining the relative effectiveness of two instructional procedures would, of course, require that individual students within classes be assessed to get individual measurements, but the individual measurements would be important only to the extent that they contributed to a single, overall summary statistic for the class, like the mean. Under H_0, each class would provide the same measurement(s) as if it had been assigned to the other instructional procedure, and so every student's score within a class (and, consequently, the mean of those scores) would be associated with the class, wherever the class was assigned.

If a complex experimental unit can be given a treatment at the same time for all subjects within the unit, as in the previous example, only one treatment time is associated with the unit. If it should be necessary to give subjects within an assignment unit a treatment at different times, then each subject should have a preassigned treatment time. For instance, every member of a family would have a treatment time fixed before the family-plus-treatment-times-for-individual-members is assigned as a unit to a particular treatment.

A second type of complex experimental unit is that in which there are elements other than subjects and treatment times incorporated into the assignment unit for the purpose of controlling for their effects by random assignment. One such element is the experimenter. In some experiments (e.g., in psychology) there is good reason to believe that different experimenters employing the same treatment would get different responses, making it necessary to control for the experimenter whenever more than one experimenter is used in a single experiment. This control can be gained by systematic association of the experimenters with the experimental units.

Example 2.1

For practical reasons it is found necessary to use two experimenters in an experiment comparing the effectiveness of treatments A and B. Four subjects are used. The four subjects are listed in alphabetical order; the first two subjects and their treatment times are associated with experimenter E_1 and the last two with experimenter E_2. There are thus four experimental units formed, each consisting of one subject, a treatment time, and the experimenter who will give the experimental treatment. The following results were obtained:

A		B	
$(S_2 - t_2 - E_1)$	13	$(S_1 - t_1 - E_1)$	5
$(S_3 - t_3 - E_2)$	17	$(S_4 - t_4 - E_2)$	8

The null hypothesis is that the measurement for each subject-treatment-time-experimenter unit is the same as it would have

been under the alternative treatment. The significance value
for t, therefore, can be determined by using the six data permutations associated with the six possible assignments of the experimental units.

Thus we can exercise statistical control over the effect of a
variable that is not of interest in an experiment by associating
a particular level of the variable systematically with an experimental unit so that, as a component of the experimental unit,
it is assigned randomly to a treatment.

Example 2.2

To determine whether a certain new drug is more effective than
a placebo, an experimenter has to use two nurses to inject
the experimental subjects. Since the appearance, attitude, and
other characteristics of the nurses may have some effect on the
response to the injection, it is decided to control for the nurse
variable. That is done in the same manner as the control for
the experimenter variable. Half of the subjects are designated
as subjects for nurse X, and half as subjects for nurse Y. The
experimental unit that is randomly assigned to drug or placebo
is a subject-treatment-time-nurse combination. One of the
experimental units might be subject 1 receiving an injection from
nurse Y at 2:15 P.M. It would be determined, in a random
manner, whether the injection would be the new drug or the
placebo; but the time and the nurse to do the administration
would not be determined randomly, because they would have
been determined systematically for subject 1 prior to the random
assignment and would be components of the experimental unit.

2.4   CONVENTIONAL TYPES OF RANDOM ASSIGNMENT

Although there are many possible types of random assignment
that can be employed in experiments, only a few types are commonly used. In this section we will discuss the two most common
random assignment procedures: those for the typical independent-groups and repeated-measures experiments. These are the
random assignment procedures which are assumed by t and F
tables in determining significance of experimental results. The
discussion will be in terms of subjects and their treatment times,
but what will be said applies equally to more complex experimental units.

Independent-groups experiments are those in which each subject takes only one treatment. One-way analysis of variance (ANOVA) and the independent t test are frequently used to analyze data from such experiments. Random assignment is conducted with sample-size constraints. If an experimenter decides to divide 10 available subjects randomly among treatments A, B, and C, with the constraint that A and B will each have three subjects and C will have four subjects, the random assignment is conducted in a different way from what it would have been for other sample-size distributions of the 10 subjects. He selects in a random manner (as by means of a lottery or a table of random numbers) three of the 10 subjects (with their treatment times) to take treatment A; three of the remaining seven are randomly selected for treatment B, and the remaining four take treatment C. There are $10!/3!3!4! = 4,200$ possible assignments. In general, for N subjects assigned to k different treatments with sample-size constraints of this kind, there are $N!/n_1!n_2!\cdots n_k!$ possible assignments, where n with a subscript indicates the sample size for the treatment indicated by the subscript.

In the standard repeated-measures experiment, each subject takes all treatments. Frequently-used statistical procedures for such experiments are the correlated t test and repeated-measures ANOVA. Suppose there are five subjects, each of whom will take treatments A, B, and C. Three treatment times are designated for the first subject, then those times are assigned randomly to the three treatments. For each subject this process is repeated, there being a designation of three treatment times, then a random determination of which treatment time to associate with a treatment. Altogether there are $(3!)^5 = 7,776$ possible assignments since there are 3! possible assignments for each subject, each of which could be associated with 3! possible assignments for each of the other subjects. In general, for N subjects each taking all k treatments, there are $(k!)^N$ possible assignments.

These two types of random assignment are the basic components of more complex forms of random assignment. An example is a treatments-by-subjects experiment, in which subjects are divided into males and females and, within each sex, are assigned randomly to treatments. Within each sex there would be $N!/n_1!n_2!\cdots n_k!$ possible assignments, where N is the total number of subjects of that sex and n refers to the number of subjects of that sex to be assigned to a particular treatment. The number of distinct assignments would be the product of the number of possible assignments for each sex, since each assignment of males could be associated with each assignment of females.

In each of the above examples of random assignment, more than one subject was assigned. It is generally assumed to be a requirement imposed by parametric significance tables. The tables are based on the assumption of random sampling from a population whose variance is estimated by the sample variance, and more than one subject would appear to be required for providing such an estimate.

Some of the random assignment conventions almost universally followed by researchers are:

1. More than one subject is assigned.
2. The number of subjects assigned to each treatment is fixed by the experimenter, not determined randomly.
3. The treatment assignment possibilities are the same for each subject:
    a. for independent-groups experiments, any subject can be assigned to any one of the treatments; and
    b. for repeated-measures experiments each subject is assigned to all treatments.

2.5   UNCONVENTIONAL TYPES OF RANDOM ASSIGNMENT

Randomization tests can be used to determine significance for experiments involving the standard independent-groups or repeated-measures random assignment procedures described in Section 2.4. If a conventional test statistic (like t or F) is computed, then a conventional test has been run, with significance determined by a randomization test procedure. But randomization tests are not restricted to such applications, because they can be used with *any* test statistic and *any* random assignment procedure, not just conventional procedures. Randomization tests based on various unconventional random assignment procedures are distributed throughout this book; these are indeed new tests, not just new ways of determining significance for old tests. By allowing the experimenter alternatives to conventional random assignment, such randomization tests provide more freedom in the conduct of an experiment.

In Section 2.4 it was indicated that it is conventional to use multiple-subject experiments because single-subject experiments are not considered valid. A randomization test, however, can give valid significance values for treatment effects for a single-subject experiment. One type of random assignment that can be used for a single-subject experiment is to decide in advance how many times the subject will take each of the treatments, and then, from the total number of treatment times required, to assign, in a random fashion, the predetermined number to each of the treatments. Significance is then determined by a randomization test in the same way as if the N measurements represented one measurement from each of N randomly assigned subjects in an independent-groups experiment. This is one of the random assignment procedures to be discussed in Chapter 10.

Another type of unconventional random assignment which will be discussed further in Chapter 10 is one in which, contrary to the conventional random assignment practice, the sample size is determined randomly rather than being fixed in advance. For instance, in a single-subject experiment with 120 successive blocks of time, one of the middle 100 blocks may be selected at random for the beginning of the block to serve as the point where a special treatment will be introduced, that treatment remaining in effect over the subsequent blocks. All blocks before the special treatment intervention are control blocks, and all blocks after the intervention are experimental blocks. The random assignment procedure randomly divides the 120 blocks of time into two sets: those for the preintervention (control) condition, and those for the postintervention (experimental) condition. The blocks are by no means independently assigned to the treatments: if a particular block is assigned to the postintervention condition, then every block after it must also be assigned to that treatment. With this special random assignment procedure the *number of treatment blocks* assigned to a treatment is *determined randomly*, and this completely determines which blocks are assigned to that treatment. This is a procedure appropriate for certain single-subject experiments, and randomization tests can permute the data according to the random assignment procedure, thereby deriving valid P-values for the treatment effect.

A conventional random assignment requirement is that treatment possibilities be the same for all subjects, but that requirement can be ignored when significance is determined by the randomization test procedure. For instance, in multiple-subject independent-groups experiments involving treatments A, B, C, and D, a restricted-alternatives assignment procedure can be used, whereby the possible alternative treatment assignments vary from one subject to the next. Some of the subjects are assigned randomly to any one of the four treatments, while others are assigned randomly within certain subsets of the four treatments. Some subjects may be willing to be assigned to A, B, or C, but not to D, so that D is not a possible assignment for them. Also, some subjects may be willing to participate in the experiment if they are assigned to A or B, but not otherwise; so for those subjects their random assignment is restricted accordingly. Restricted-alternatives random assignment is feasible also for repeated-measures experiments. Some subjects may take all four treatments while others take only two or three, and random assignment of treatment times to treatments for each subject is made according to the alternatives available to that subject. The use of randomization tests for restricted-alternatives random assignment is discussed in Chapter 6.

3

Calculating Significance Values

3.1 INTRODUCTION

Even with fairly small sample sizes the number of data permutations required makes randomization tests impractical unless a computer is used. This chapter is concerned with calculating significance values when randomization tests are performed by a computer, but first it will be necessary to consider calculation for randomization tests in more detail to determine what it is that the computer must do.

Example 3.1

Four subjects are selected for an experiment. Each experimental unit is a subject plus a predetermined treatment time. The letters a, b, c, and d will be used to indicate the four subjects and their treatment times. Use of a random procedure leads to the selection of experimental units a and c for treatment A; the remaining two units, b and d, take treatment B. The following results are obtained and t is computed by use of formula (1.1), where $\overline{X}_A$ was predicted to be larger than $\overline{X}_B$:

A		B	
Unit	Measurement	Unit	Measurement
a	6	b	3

A			B	
Unit	Measurement		Unit	Measurement
c	7		d	4
$\overline{X} = 6.5$			$\overline{X} = 3.5$	
		$t = 4.24$		

The random assignment procedure determines the equally probable assignments. There are 4!/2!2! = 6 of them. H_0 of
identical treatment effects implies that wherever a unit is assigned, its measurement goes with it. Under H_0 there is thus a
data permutation associated with each possible subject assignment, providing the following six data permutations:

A	B	A	B	A	B	A	B	A	B	A	B
3	6	3	4	3	4	4	3	4	3	6	3
4	7	6	7	7	6	6	7	7	6	7	4
$\overline{X} = 3.5$	6.5	4.5	5.5	5	5	5	5	5.5	4.5	6.5	3.5
$t = -4.24$		-0.47		0		0		0.47		4.24	

For a test where it had been predicted that $\overline{X}_A$ would be
greater than $\overline{X}_B$, the proportion of the data permutations having
as large a value of t as 4.24 would be determined. The data
permutation representing the obtained results is the only one
meeting the requirement, so the proportion is 1/6. The one-
tailed significance or P-value associated with the obtained results
is 1/6, or about 0.167.

If there had been no prediction of the treatment that would
provide the larger mean, a two-tailed test would have been
appropriate. The proportion of the data permutations with as
large a value of $|t|$ as 4.24, the obtained value, is determined.
This proportion is 2/6, or about 0.333. The significance or
P-value associated with a two-tailed test is thus about 0.333.

Use of the t table to determine significance for the one-tailed
test would have suggested that H_0 could be rejected at the 0.05
level of significance. The use of t tables with such small samples is inappropriate and would be unacceptable to most researchers.

Example 3.2

There are two subjects in an experiment in which each subject takes both experimental treatments. Each subject has two treatment times, making four experimental units. The designations a_1, a_2, b_1, and b_2 will be used to specify the two subjects and their treatment times, where subscript "2" refers to the later treatment time for a subject. Treatment A is expected to provide the larger mean. It is randomly determined (as by tossing a coin) which treatment time for subject a is for treatment A and which for treatment B. Then it is determined randomly for subject b which treatment time goes with each treatment. The random assignment leads to subject a taking treatment B first and subject b taking treatment A first, with the following results:

A		B		$X_A - X_B$
Unit	Measurement	Unit	Measurement	(D)
a_2	10	a_1	8	2
b_1	7	b_2	6	1
	$\bar{X} = 8.5$		$\bar{X} = 7$	$\bar{D} = 1.5$

(If B had been predicted to provide the larger mean, D would have been $X_B - X_A$.) Because each subject takes both treatments, the test statistic t is computed from the values of D (the difference scores in the last column), using the following formula for correlated t, where N is the number of difference scores:

$$t = \frac{\bar{D}}{[\Sigma(D - \bar{D})^2/(N^2 - N)]^{1/2}} \qquad (3.1)$$

Formula (3.1) gives a t value of 3 for the above results. There were two possible assignments for subject a, each of which could be associated with either of two possible assignments for subject b, making $2 \times 2 = 4$ possible assignments. Under H_0 the measurement for any experimental unit is the same as it would have been if that unit had been assigned to the other treatment; in other words, at each treatment time a subject's response was the same as it would have been at that treatment time under the other treatment. There are thus four data permutations that can be specified for the four possible assignments. The four data permutations are:

A	B	A	B	A	B	A	B
8	10	8	10	10	8	10	8
6	7	7	6	6	7	7	6
$\bar{X} = 7$	8.5	7.5	8	8	7.5	8.5	7
t =	−3		−0.33		0.33		3

Although there are the same number of experimental units as in Example 3.1, there are only four possible assignments because of the random assignment procedure, and consequently only four data permutations instead of six. The following permutation, for example, is *not* used:

A	B
8	6
10	7

It is not used because it would be associated with an impossible assignment, namely the assignment where subject a took treatment A at both of a's treatment times and subject b took treatment B at both of b's treatment times. Since the random assignment procedure required each subject to take both treatments, the only data permutations to be used in testing H_0 are those where, for each subject, one of the two obtained measurements is allocated to one treatment and the other measurement to the other treatment.

Inasmuch as the obtained results (represented by the last of the four permutations) show the largest t value, a one-tailed test where the direction of difference was correctly predicted would give a P-value of 1/4, or 0.25.

3.2 CRITERIA OF VALIDITY FOR RANDOMIZATION TESTS

The validity of randomization tests for various types of application has been pointed out earlier. It must be stressed, however, that the use of a randomization test procedure does not guarantee validity. A randomization test is valid only if it is properly conducted. In light of the numerous test statistics and random assignment procedures that can be used with randomization tests, it is essential for the experimenter to know basic rules for the valid execution of a randomization test.

Before dealing with the validity of randomization test procedures, it will be useful to specify criteria of validity for statistical testing procedures in general. Within the decision-theory model of hypothesis testing, which requires a researcher to set a level of significance in advance of the research, the following criterion is appropriate:

Decision-theory Validity Criterion

A statistical testing procedure is valid if the probability of a type I error (rejecting H_0 when true) is no greater than α, the level of significance, for any α.

For instance, the practice of conducting a one-tailed t test in accord with the *obtained*, rather than the predicted, direction of difference between the means is an invalid procedure because the probability of rejecting H_0 when it is true is greater than α.

The above criterion, which is implicit in most discussions of the validity of statistical testing procedures, unfortunately is expressed in terms associated with the decision-theory model: "rejection," "type I error," and "α." This may create the impression that only within the Neyman-Pearson decision-theory framework of hypothesis testing can one have a valid test. Restriction of validity to situations with a fixed level of significance may be satisfactory for quality control in industry, but it is unsatisfactory for scientific experimentation. Interest in levels of significance set in advance is by no means universal. For the numerous experimenters who are interested in using the smallness of a P-value as an indication of the strength of evidence against H_0 (an interpretation of P values inconsistent with the decision-theory approach), a more general validity criterion is required. An *operationally equivalent* validity criterion that does not use decision-theory terminology (Edgington, 1970) follows:

General Validity Criterion

A statistical testing procedure is valid if, under the null hypothesis, the probability of an exact probability or significance value as small as P is no greater than P, for any P.

For instance, under the null hypothesis the probability of obtaining a significance value (P-value) as small as 0.05 must be no greater than 0.05, obtaining one as small as 0.03 must be no greater than 0.03, and so on. Obviously, the two criteria of validity are equivalent: for any procedure they lead to the same conclusion regarding validity. The general validity criterion, therefore, can be used by experimenters interested in the decision-theory approach. It is useful also to experimenters who do not set levels of significance in advance but instead use the smallness of the P-value as an indication of the strength of the evidence against H_0. (Such experimenters may, of course, *report* their results as significant at the smallest conventional α level permitted by their results.) When a statistical testing procedure is valid according to the general validity criterion, the P-value itself is the probability of getting such a small P-value when H_0 is true.

It is useful to know the conditions that must be satisfied for certain testing procedures to meet the general validity criterion. Depending on the testing procedure, the conditions or assumptions may include random sampling, random assignment, normality, homogeneity of variance, homoscedasticity, or various other conditions. In the following discussion, we will consider components of a valid *randomization test* procedure.

Instead of describing the assumptions necessary for a valid randomization test, we will specify steps that are sufficient to ensure the validity of a randomization test procedure for testing the following null hypothesis:

Randomization Test Null Hypothesis

The measurement (or set of measurements) associated with each experimental unit is independent of the assignment of units to treatments.

The steps to be taken are:

Step 1. Specify the test statistic before the experiment, and ensure that the test statistic is defined in such a way that it can be computed for a data permutation without considering whether the data permutation represents the obtained results.

Step 2. List every equally probable assignment of experimental units to treatments. (This step presupposes a random assignment to treatments.)

Step 3. Within each of the equally probable assignments, substitute for each experimental unit the measurement (or set of measurements) for the unit to provide a distribution of data permutations.

Step 4. For each data permutation compute the test statistic specified in step 1.

Step 5. Determine the proportion of the data permutations with as large a test statistic value as the obtained test statistic value, and use that proportion as the P-value.

Let us now illustrate these steps with a numerical example.

Example 3.3

Step 1. T_A, the total of the measurements for treatment A, is specified as the test statistic.

Step 2. Two of four subjects are to be selected randomly to take treatment A, and the remaining two take treatment B. The six equally probable assignments that can be made are: (1) A: a, b; B: c, d; (2) A: a, c; B: b, d; (3) A: a, d; B: b, c; (4) A: b, c; B: a, d; (5) A: b, d; B: a, c; (6) A: c, d; B: a, b.

Step 3. Subjects a, b, c, and d have measurements 5, 8, 3, and 4, respectively. Thus the data permutations corresponding to the previous six assignments, listed in the same order as the assignments, are: (1) A: 5, 8; B: 3, 4; (2) A: 5, 3; B: 8, 4; (3) A: 5, 4; B: 8, 3; (4) A: 8, 3; B: 5, 4; (5) A: 8, 4; B: 5, 3; (6) A: 3, 4; B: 5, 8.

Step 4. The values of T_A, the test statistic, for the data permutations are: (1) 13; (2) 8; (3) 9; (4) 11; (5) 12; (6) 7.

Step 5. An obtained test statistic value must be one of the six values given by step 4. The P-value assigned to an obtained test statistic value is the proportion of the values listed in step 4 that are as large as the obtained value.

Although this is a numerical example, there is no specification of the experimental results. Given that H_0 is true and that steps 1 to 4 are carried out, the obtained test statistic value for step 5 is a random variable that can, with equal probability, take on any of the six values given in step 4.

Thus the probability of obtaining a test statistic value that is the largest (and consequently assigning a P-value of 1/6 to the obtained results) is 1/6, the probability of obtaining a value that is one of the two largest (and consequently getting a P-value of 2/6) is 2/6, and so on. Thus the procedure of determining significance by following the five steps is valid. (If the example had involved tied test statistic values, the probability of getting a significance value as small as P might be smaller than P, but never larger. The reasoning to be given in Section 3.4 regarding tied statistic values is applicable here.)

If the test statistic for step 1 in this example had not been specified in such a way that it could be computed for a data permutation without considering whether that data permutation represented the obtained results, the procedure of determining significance would have been invalid. For example, use of "$\bar{X}$ for the treatment which provides the larger $\bar{X}$ for the obtained results" would have been unacceptable for this example, because there would be no unique list of test statistic values for step 4. If assignment 5 provided the experimental results, $\bar{X}_A$ for all data permutations would provide the distribution of test statistic values for step 4. On the other hand, if assignment 2 or certain other assignments had been made, the test statistic would have been $\bar{X}_B$ for all data permutations. The theoretical distribution of test statistic values for step 4 must not depend on which data permutation represents the obtained results. Only if the values in step 4 are independent of the obtained results (the obtained data permutation) can they be regarded, under H_0, as equally probable values for a random variable.

The preceding discussion assumes that the experimenter expects a treatment effect to lead to a larger value of the test statistic than if there were no treatment effect. Sometimes it is simpler to use a test statistic whose *smallness* would be expected to reflect the treatment effect. In such cases, of course, the P-value is the proportion of the data permutations providing such a small test statistic value as the obtained value. Whether the P-value is to be based on the smallness or the largeness of the obtained test statistic value should be specified in step 1.

The five steps that have been discussed refer to the application of a randomization test where all possible assignments are considered. It is possible, however, to have a valid application of a randomization test based on a random sample of all possible assignments. That type of application involves the

random data permutation procedure that will be described in
Section 3.4. When the random data permutation procedure is
followed, step 2 should be changed to: "Take a random sample
of r random assignments, and list those r assignments plus the
actual assignment as a distribution of equally probable assignments." If this change is made, following all five steps will
ensure the validity of application of the random data permutation procedure.

3.3   PERMUTING DATA FOR EXPERIMENTS WITH
         EQUAL SAMPLE SIZES

There are special techniques for permuting data that can be
used with equal sample sizes, techniques that use only a subset of the data permutations associated with all possible assignments. In earlier numerical examples involving the t test with
equal sample sizes, there is an even number of data permutations
and they can be paired so that one member of a pair is the
mirror image of the other member in the sense that the measurements are divided into two groups in the same way with the
treatment designations reversed. In Example 3.1 the first permutation was: A: 3, 4; B: 6, 7, while the last permutation
was: A: 6, 7; B: 3, 4. Obviously, when there are equal
sample sizes for two treatments, all data permutations can be
paired in this fashion; if there was one for which there was
no such mirror image, it would indicate that a possible assign-
ment had been overlooked. Since $|t|$ only depends on which
measurements are in each of the two groups and not at all on
the treatment designations associated with the two groups of
measurements, both members of each mirror-image pair must
give the same value of $|t|$. If we divide all data permutations
into two subsets, where each subset contains one member of a
mirror-image pair, then the proportion of t statistics that have
an absolute value as large as some specified value is the same
for both subsets. Consequently, the proportion of t statistics,
in either of the subsets, that have an absolute value as large as
some value is the same as the proportion for the entire set.
If significance was determined by using one of the two subsets,
the same two-tailed P-value for t would be obtained as if the
entire set of data permutations was used. The amount of computation, however, would be cut in half.

   Mirroring of the same kind occurs with ANOVA, which can
deal with more than two treatments. Consider the division of

six measurements into three groups: 2, 4; 7, 3; 12, 14. If
there was a data permutation where that division of measurements represented the measurements for treatments A, B, and
C, respectively, there would have to be five other permutations
for the same division of measurements, corresponding to the
other five ways in which the treatment designations could be
assigned to those three groups. Of course, all six data per-
mutations would give the same value of F because F, like $|t|$,
is a nondirectional test statistic. Thus, by taking only one-
sixth of the set of permutations, one could determine the P-value
for F to get the same value as with the entire set. For k
treatments with equal sample sizes, a subset of 1/k! of all permutations with each data division represented only once in the
subset (not several times with different treatment designations)
is all that is necessary to determine the P-value that would be
obtained by using the entire set of data permutations. This
can, of course, result in a considerable saving of computational
time.

3.4   RANDOM DATA PERMUTATION

There are two basic methods of permuting data to get significance by the use of a randomization test. The naming of these
two methods has not been standardized, but we will call them
systematic data permutation and *random data permutation*. The
numerical examples that have been given have employed the
systematic method, where data permutations in a set or subset
are permuted systematically (nonrandomly) in determining significance.

   Random data permutation is a method that uses a random
sample of all possible data permutations to determine significance.
(Alternatively, random data permutation can be regarded as a
method that uses data permutations associated with a random
sample of possible assignments.) It serves the same function
as systematic data permutation with a substantial reduction in
the number of permutations that need to be considered. Instead
of requiring millions or billions of data permutations, as would
be required for the systematic data permutation method for many
applications of randomization tests, the random data permutation
method may be effective with as few as 1,000 data permutations.
The following discussion is based on Edgington (1969a, pp.
152—155; 1969b). The procedure for determining significance
is the same as that of Dwass (1957) and Hope (1968), but the

power of the random data permutation procedure is presented
in a different way.

Suppose the experimenter decides that the time and money
to deal with all data permutations in the relevant set or subset
is too great to be practical. He can then proceed in the following way, using the random permutation method. He performs
999 random data permutations, which is equivalent to selecting
999 data permutations at random from the set that would be used
with the systematic method. Under H_0, the data permutation
representing the obtained results is also selected randomly from
the same set. Thus, given the truth of H_0, we have 1,000
data permutations that have been selected randomly from the
same set, one of which represents the obtained results. We
determine the significance as the proportion of the 1,000 test
statistic values that are as large as the obtained value. That
this is a valid procedure for determining significance can be
readily demonstrated. Under H_0, the obtained test statistic
value (like the 999 associated with the 999 data permutations
selected at random) can be regarded as randomly selected from
the set of all possible test statistic values, so that we have
1,000 randomly selected values, one of which is the obtained
value. If all 1,000 test statistic values were different values,
so that they could be ranked from low to high with no ties,
under H_0 the probability would be 1/1,000, or 0.001, that the
obtained test statistic value would have any specified rank from
1 to 1,000. So the probability that the obtained test statistic
value would be the largest of the 1,000 would be 0.001. If
some values were identical, the probability could be less than
0.001, but no greater. Admitting the possibility of ties, the
probability is *no greater than* 0.001 that the obtained test
statistic value would be larger than all of the other 999 values.
Similarly, the probability is no greater than 0.002 that it would
be one of the two largest of the 1,000 values, and so on. In
general, when H_0 is true, the probability of getting a P-value
as small as P is no greater than P, and so the method is valid.

A randomization test using the random data permutation
method, although employing only a sample of the possible data
permutations, is therefore valid. When H_0 is true, the probability of rejecting it at any α level is no greater than α; but
if H_0 is false and there is an actual treatment effect, random
data permutation is less powerful than systematic data permutation based on the entire set of possible assignments. Increasing the number of data permutations used with the random data

permutation method increases the power of a randomization test employing random data permutation.

The computer programs in this book for random data permutation permit the experimenter to specify the number of data permutations to be performed. If he specifies N permutations the computer will treat the obtained results as the first data permutation and randomly permute the data an additional N − 1 times. Using 1,000 data permutations does not provide the power given by several thousand, but the power is still substantial. For example, the probability is 0.99 that an obtained test statistic value that would be judged significant at the 0.01 level, by using systematic data permutation, will be given a P-value no greater than 0.018 by random data permutation with 1,000 random permutations. Also, the probability is 0.99 that an obtained test statistic value that would be found significant at the 0.05 level, according to systematic data permutation, will be given a P-value no greater than 0.066 by random data permutation using 1,000 random permutations.

Because the systematic and random data permutation methods permute the data differently, different computer programs are required. The distinction between systematic and random data permutation, however, should be recognized as a technical one that has practical importance for performing a randomization test but does not affect the interpretation of a significance value given by a randomization test. Nevertheless, some people believe that the distinction between computing P-values on the basis of systematic and random data permutation is sufficiently important to merit specifying in a research report which procedure was used to determine significance. As the power of a random data permutation procedure relative to a systematic procedure generating all relevant data permutations is a function of the number of random data permutations employed, that number should be provided in a report of results of a randomization test whenever a random permutation procedure is used.

3.5 EQUIVALENT TEST STATISTICS

Examination of the data permutations for any of the t tests in this chapter or in Chapter 1 shows that, within each set of data permutations, the largest $|t|$ is associated with the data permutation with the largest $|\bar{X}_A - \bar{X}_B|$, and the smallest $|t|$ is associated with the data permutation having the smallest

$|\bar{X}_A - \bar{X}_B|$. In fact, if data permutations for t tests are ranked from high to low with respect to $|t|$, they always will be found to be ranked from high to low also with respect to $|\bar{X}_A - \bar{X}_B|$. Consequently, the proportion of the data permutations with as large a value of $|t|$ as the obtained value is the proportion with as large a value of $|\bar{X}_A - \bar{X}_B|$ as the obtained value. Thus, $|t|$ and $|\bar{X}_A - \bar{X}_B|$ are two different test statistics which give same P-value for a randomization test. Therefore, one could use the simpler test statistic $|\bar{X}_A - \bar{X}_B|$ to determine significance and associate the P-value with $|t|$ for the obtained results.

Two test statistics which must give the same P-value for a randomization test are called *equivalent test statistics*. It can save time to compute a simpler, equivalent test statistic to t, F, r, or some other conventional test statistic for every data permutation to determine the significance value for the more complex one. Two test statistics are equivalent if and only if they are perfectly monotonically correlated over all data permutations in the set. Expressed in terms of correlation, there will be a perfect positive or negative rank correlation between the values of the two test statistics over the set of data permutations used for determining significance. In the following chapters, equivalent test statistics for a number of common test statistics will be given.

Most of the examples of derivations of simpler equivalent test statistics that will be given in later chapters involve several steps. The first step is to simplify a conventional parametric test statistic somewhat, and each successive step simplifies the test statistic resulting from the preceding step. Various techniques, resulting both in linearly and nonlinearly related equivalent test statistics, will be employed in those chapters.

The test statistic for a randomization test can be simpler because the test statistic needs to reflect only the type of treatment effect that is anticipated, unlike parametric test statistics, which also must involve an estimate of the variability of the component reflecting the anticipated effect. For example, the numerator of t, which is the difference between means, is what is sensitive to a treatment effect; the denominator is necessary for estimating the variability of the numerator. For other parametric test statistics as well, the denominator very frequently serves the function of estimating the variability of the numerator, under the null hypothesis, whereas the randomization test procedure generates its own null distribution of the measure of effect, such as a difference between means, making the denominator irrelevant.

3.6 RANDOMIZATION TESTS IN THE COMPUTER AGE

The amount of computation required for randomization tests made them impractical in the days when computation had to be done by hand or by mechanical desk calculator. The number of data permutations required is large even for relatively small samples, if systematic data permutation is employed. For example, an independent t test with only seven subjects in each of two groups requires consideration of $14!/7!7! = 3,432$ possible assignments, of which half (i.e., 1,716) must be used for determination of significance. Because a t value or a value of an equivalent test statistic must be computed for each of the 1,716 permutations, the computational requirements are considerable. Thus, even for small samples, the amount of computation for a randomization test is great. However, although several thousand computations of a test statistic is an excessive number when the computations are done by primitive methods, in the modern age of computers a few thousand computations is not at all expensive or time consuming. Furthermore, the possibility of using random data permutation to base a randomization test on a random sample of all possible data permutations makes randomization tests practical even for large samples.

Table 1 shows computer times and costs for several applications of randomization tests, determined in 1985 to update a similar study by Edgington and Strain (1973). The sample sizes in the two studies are not all the same, but where they are the 1985 times and costs, based on use of a Honeywell Multics DPS8—MR10.2 computer, range from one-fifth to one-half of the 1973 times and costs, which were for a CDC 6400 computer. The two studies used the same computer programs, namely, Programs 4.2, 4.3, 5.1, and 5.2, given in Chapters 4 and 5. The systematic programs are for equal-N designs, and the random programs used 1,000 permutations. Although times and costs will vary over computer installations, Table 1 is instructive in showing how various factors influence computer time for randomization tests.

Notice that for a particular type of test, using systematic data permutation, the time is proportional to the number of data permutations. For example, there are about four times as many data permutations for an independent t test with ten subjects as with nine subjects, and the time is four times as great. Similarly, for correlated t tests, using systematic permutation, there are twice as many data permutations for 16 subjects as for 15, and the time is twice as large.

TABLE 1   Examples of Computer Times and Costs for Randomization Tests

Type of test	Data permutation method	No. of treat-ments	Subjects per treatment	Time (sec)	Cost* ($)
Indep. t	Systematic	2	9	5	0.25
Indep. t	Systematic	2	10	20	0.68
Indep. t	Random	2	10	1	0.06
Indep. t	Random	2	50	5	0.25
Corr. t	Systematic	2	15	6	0.24
Corr. t	Systematic	2	16	12	0.43
Corr. t	Random	2	25	3	0.09
Corr. t	Random	2	50	6	0.24
One-way ANOVA	Systematic	4	3	4	0.17
One-way ANOVA	Systematic	4	4	658	22.18
One-way ANOVA	Random	5	10	4	0.18
One-way ANOVA	Random	5	20	8	0.32
Repeated-measures	Systematic	4	4	5	0.20
Repeated-measures	Systematic	4	5	135	4.55
Repeated-meausres	Random	5	10	5	0.19
Repeated-measures	Random	10	10	9	0.36

*Canadian currency.

For the random permutation procedure the number of data permutations is the same (i.e., 1000) for all sample sizes, and it will be noted that the time for random programs for a particular type of test is proportional to the sample size (number of data points). The same relationship holds for systematic data permutation, but there the number of data permutations increases so rapidly with an increase in sample size that it masks the fact that time is proportional to sample size.

The number of data permutations for a systematic program mounts rapidly with an increase in sample size. For one-way ANOVA, with 3 subjects in each of four groups, Table 1 shows that the time to compute the P-value is 4 seconds and that the cost is 17 cents. The addition of only one more subject per group, however, brings the time up to 658 seconds, for a cost of $22.18. For repeated-measures ANOVA, with 4 subjects each taking 4 treatments, the computer time is 5 seconds and the cost is 20 cents, but the addition of a fifth subject increases the time to 135 seconds and the cost to $4.55. Imagine an unsuspecting researcher using a one-way ANOVA systematic program with 4 groups of 5 subjects each; extrapolating from the $22.18 for four groups of 4 subjects each, we see that it would cost over $3,000 for the test. In order to avoid excessive computing costs, therefore, it is important to know in advance of using a systematic permutation program just how many data permutations would be generated. It is useful to have the computer perform computation of the number of data permutations as soon as the data are entered and to have the computer programmed to switch to a random permutation procedure automatically if the number of data permutations exceeds some user-specified value. (NOTE: The "skeleton" programs in this book do *not* have such a built-in safeguard, so use them with care until a safeguard is incorporated.)

3.7 WRITING PROGRAMS FOR RANDOMIZATION TESTS

For many applications of randomization tests the experimenter can use programs contained in this book. The language used in the programs is Fortran IV, a language commonly used for statistics; but the programs can be easily translated into another computer language. For either rewriting the programs contained in the book or writing new programs, it is necessary to know what the programs require the computer to do.

All computer programs for randomization tests must specify the performance of the following operations:

1. Compute an obtained test statistic value, which is a test statistic value for the results given by the experiment.
2. Systematically or randomly permute the data.
3. Compute the test statistic value for each data permutation.
4. Increment a counter each time a test statistic value is as large as the obtained value or, where a treatment effect is manifested by the smallness of the test statistic, each time a test statistic is as small as the obtained value.
5. Compute the significance or probability value. The probability value is the proportion of the test statistic values, including the obtained value, that are as large as (or, where appropriate, as small as) the obtained test statistic value.

Figure 1 shows the basic structure of the computer programs given in later chapters. In this flow chart, as in the programs, the details of INPUT and OUTPUT are omitted because their general nature can be inferred from the body of the program.

For systematic data permutation it is essential that each data permutation in the set (or subset) of data permutations associated with the relevant possible assignments be determined. The problem faced by the computer in making certain that all data permutations have been considered once and only once is that which confronts the experimenter if data permutations are done by hand. Imagine the difficulty in checking several thousand data permutations to see if all possible permutations are represented. Obviously some sort of systematic listing procedure is required. In a program using systematic data permutation, the counterpart of systematic listing is specification of a systematic sequence for generating data permutations. Different types of random assignment require different sets of data permutations, and so there are correspondingly different procedures for systematically producing the data permutations. For each type of random assignment it must be determined how one can systematically produce the possible permutations.

Example 3.4

We have five subjects, two of which are randomly assigned to A and three to B. There will be $5!/2!3! = 10$ data permutations.

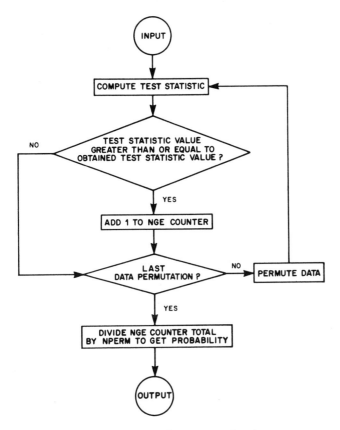

FIGURE 1   Flow chart for randomization tests.

We obtain the following results: A: 12, 10; B: 9, 12, 8.
How can we list the 10 data permutations systematically to make
sure each one is listed once and only once? We want a procedure that can be generalized to cases where there are so many
data permutations that a definite sequential procedure of listing
the data permutations is the only way to be certain that the
list is correct. When there are two treatments, all we need is
a systematic listing procedure for the measurements allocated
to one of the treatments, because the remaining measurements
are allocated automatically to the other treatment. Thus what
is needed is a listing procedure for the 10 possible pairs of
measurements for treatment A. Such a procedure results from

associating an index number with each of the five measurements, systematically listing the pairs of index numbers, then translating the index numbers back into the measurements with which they are associated. We will use the integers 1 to 5 as index numbers representing the measurements in the order in which they were listed

Measurements	12	10	9	12	8
Index numbers	1	2	3	4	5

The list of all possible pairs of index numbers that can be associated with the measurements allocated to treatment A for the 10 data permutations is: (1,2), (1,3), (1,4), (1,5), (2,3), (2,4), (2,5), (3,4), (3,5), (4,5). These index numbers are in ascending order within the two values for each permutation, then the permutations of index numbers are listed in ascending order, treating each permutation as a two-digit number. This procedure pairs each index number with every other one and eliminates the possibility of listing the same combination of index numbers with the index numbers in two different orders. Translating the index number list into a list of measurements for treatment A produces the following: (12, 10), (12, 9), (12, 12), (12, 8), (10, 9), (10, 12), (10, 8), (9, 12), (9, 8), (12, 8). For each of these 10 pairs of measurements for A, the remaining three measurements are associated with the B treatment, providing the 10 data permutations.

Earlier in this chapter it was pointed out that when there are equal sample sizes for t tests or ANOVA, it is necessary to compute only 1/k! of the possible data permutations, where k is the number of treatments, because of the redundancy due to mirror-image data permutations that have the same divisions of the data but transposed treatment designations. Listing procedures have been incorporated into the equal-N programs to provide the required selection of data permutations.

When random data permutation is employed, it is not necessary to work up a systematic listing procedure, because only a sample rather than all of the data permutations are generated. The random data permutations in the programs are produced in the following manner. The original data configuration is permuted randomly to generate the first of the *randomly* selected data permutations, then that data permutation is permuted randomly to generate the second, and so on, each of the random data permutations being generated randomly from the previous data permutation.

The programs given in this book will handle a large number of applications, but effective utilization of the potential of randomization tests can only be achieved if experimenters make additional applications requiring new programs. All programs can follow the basic format given in Figure 1, so that with a slight modification a program can be changed into a new one. If the determination of significance by a randomization test is the only acceptable procedure, or if use is made of a new test for which there is no alternative method of determining significance, the expense of modifying a program or writing an entirely new one is justifiable even if the new program is used only once. However, many new programs can be employed repeatedly, either by the same experimenter or by others.

A computer program created for one test can be changed to be used for a quite different test simply by modifying the test statistic and portions of the program for computing the test statistic. For example, we have a computer program with independent t as the test statistic for detecting a treatment effect where one set of measurements is systematically larger than the other. With a simple change in the test statistic we can produce a program for detecting a difference in variability. One such test statistic would be the ratio of the within-treatment variances. To determine the significance of this new test statistic, its value would be computed for the experimentally obtained results and for each of the data permutations that otherwise would be used for determining the significance of t. The data permutation procedure must be based on the random assignment procedure used by the experimenter, but the test statistic in a program for a certain kind of random assignment can be any test statistic whatsoever, sensitive to any type of effect.

Computer programs for determining significance by data permutation for conventional experimental designs fall into two major classes: programs for experiments with independent groups where each experimental unit could be assigned to any treatment and programs for experiments of a repeated-measures kind where each experimental unit is subjected to all treatments. The two types of random assignment account for the bulk of random assignment that is performed in experiments and, since all programs for tests based on the same random assignment procedure differ only with regard to the test statistic, frequently there is need for only two basic programs, each of which can be employed with a number of different test statistics.

Other types of random assignment sometimes are used which require special types of programs and special procedures for permuting the data; but for each random assignment procedure a single basic program is sufficient for many applications, modifications in the program being made through changes in the test statistic.

3.8   HOW TO TEST SYSTEMATIC DATA PERMUTATION PROGRAMS

After a systematic data permutation program has been written, it should be checked to see if it gives correct P-values. Data sets for which the significance can be determined without the use of a computer program can be used to test programs.

If small sets of data are employed, the data permutations can easily be listed and the significance determined readily without a computer program. A computer program for systematic data permutation will give the same value if the program is written correctly.

Large sets of data also can be used to test programs for systematic data permutation. The data may be allocated to treatments in such a way as to provide the maximum value of the test statistic, and the probability value then would be the number of permutations giving the maximum value divided by the total number of permutations. For example, with completely distinctive, nonoverlapping sets of measurements with five measurements in each group, the two-tailed P-value for independent t should be $2/252$ (about 0.008), since two permutations give the maximum value and there are $10!/5!5! = 252$ data permutations. Data configurations that provide the next-to-maximum value of a test statistic also are convenient to use to test programs. Accompanying the systematic data permutation programs in the following chapters are sample data that can be used to test the programs. All or some of the data permutations and test statistic values should be printed out as a double check on the program.

3.9   HOW TO TEST RANDOM DATA PERMUTATION PROGRAMS

The correctness of a random data permutation program is harder to assess because of its random element. Rarely will a random sample of the relevant data permutations yield exactly the same P-value as systematic data permutation, but the correspondence should be close. It has been shown elsewhere that the closeness of correspondence can be expressed probabilistically (Edgington, 1969a, pp. 151—155). Given a systematic permutation probability value for a data set, and the number of permutations used for random data permutation, the probability that the value given by random data permutation will be within a certain distance of the systematic permutation probability value can be determined. This fact can be used to test random data permutation programs. The larger the number of data permutations used for random data permutation, the greater the probability that the obtained probability value will be within a certain distance of the value given by systematic data permutation; so it is useful in testing random data permutation programs to use a large number of data permutations.

Table 2 is based on 10,000 permutations. The formula for the lower limit of the 99% interval in Table 2 is

$$\frac{9,999p - 2.58(9,999pq)^{1/2} + 1}{10,000}$$

and the formula for the upper limit is

$$\frac{9,999p + 2.58(9,999pq)^{1/2} + 1}{10,000}$$

where p is the systematic permutation probability value and q is 1 — p. A new random permutation program can be tested in the following way:

1. Determine the systematic permutation P-value associated with a set of test data. (Some test data and the corresponding systematic permutation probability value are given with each of the programs in this book.)
2. Determine the P-value for the test data with the random data permutation program to be tested, using 10,000 data permutations.
3. Use Table 2 to see whether the P-value given by random permutation falls within the 99% probability interval associated with the systematic permutation P-value.
4. Repeat the above procedure with new data if the first set of data gives a P-value outside the 99% probability interval. If the probability value again falls outside the 99% probability interval, the program should be examined very carefully.

TABLE 2   99% Probability Intervals for Random Data
Permutation P-values Based on 10,000 Data Permutations

Systematic data permutation P-value	99% Probability interval for random data permutation
0.01	0.0075−0.0127
0.02	0.0165−0.0237
0.03	0.0257−0.0345
0.04	0.0350−0.0452
0.05	0.0445−0.0557
0.06	0.0540−0.0662
0.07	0.0635−0.0767
0.08	0.0731−0.0871
0.09	0.0827−0.0975
0.10	0.0923−0.1078
0.25	0.2389−0.2612
0.50	0.4871−0.5130

Table 2 is based on 10,000 data permutations, in order to give probability intervals small enough to be effective in testing new programs. Once the program has been checked out it may, of course, be used with fewer or more than 10,000 data permutations if the experimenter desires.

As with the testing of systematic permutation programs, it is desirable to double-check a random permutation program by having all or some of the data permutations and their test statistic values printed out. For this check, the number of random permutations for conducting the test could be 20 or 30, instead of 10,000.

REFERENCES

Dwass, M. (1957). Modified randomization tests for nonparametric hypotheses. *Ann. Math. Stat. 28*, 181−187.

Edgington, E. S. (1969a). *Statistical Inference: The Distribution-free Approach.* McGraw-Hill, New York.

Edgington, E. S. (1969b). Approximate randomization tests. *J. Psychol.* 72, 143−149.

Edgington, E. S. (1970). Hypothesis testing without fixed levels of significance. *J. Psychol.* 76, 109−115.

Edgington, E. S., and Strain, A. R. (1973). Randomization tests: computer time requirements. *J. Psychol.* 85, 89−95.

Hope, A. C. A. (1968). A simplified Monte Carlo significance test procedure. *J. R. Stat. Soc.* 30, 582−598.

4

One-way Analysis of Variance and the Independent t Test

4.1 INTRODUCTION

This is the first of several chapters dealing with applications of randomization tests. These chapters serve a dual function. The primary function is to illustrate randomization test principles by specific instances. A second function is to provide applications useful to experimenters. To help ensure the practical value of the applications, computer programs accompany many of the discussions.

The randomization test procedure is best understood if it is first considered in relation to familiar statistical tests. Therefore the early chapters concern applications to common analysis of variance (ANOVA) and t tests, providing the foundation for later applications to more complex and less familiar tests.

When the question of violations of parametric assumptions arises, determination of significance by means of a randomization test is a ready answer. That applies to the simple tests considered in this chapter as well as to the more complex tests in some of the later chapters. It is true that in the vast majority of applications of ANOVA and independent t tests the use of F or t tables for determining significance goes unchallenged; other researchers and editors are not likely to question the tenability of the parametric assumptions. (As the practicality of randomization tests becomes more widely known, however, this attitude may change.) However, although the *proportion* of objections to the use of the parametric significance tables is small, these tables are used so frequently that even a small proportion of objections can be a considerable number. The experimenter who applies these tests to dichotomous data, for example, is likely to encounter the objection that the data must be taken from a continuum. If the sample sizes are too small he or she also may run into objections. In such cases, the determination of significance by data permutation, using programs in this chapter, is a solution to the difficulty. Even inequality of sample sizes can cause problems. Studies of "robustness" have examined the validity of F and t tables when the parametric assumptions underlying the tables are not met, and they have shown repeatedly that parametric assumptions are important considerations when sample sizes are unequal. After discussing ANOVA with unequal sample sizes, Keppel (1973, p. 352) stated:

> This analysis should entail an increased concern for violations of the assumptions underlying the analysis of variance. As we have seen, the F test is relatively insensitive to violations of these assumptions, but only when the sample sizes are equal. Serious distortions may appear when these violations occur in experiments with unequal sample sizes.

This view expressed by Keppel is consistent with that of many researchers (and editors); so the researcher would be well advised to use equal sample sizes when possible and to use a randomization test to determine significance when he must use unequal sample sizes.

For one-way ANOVA and the independent t test, the random assignment to treatments is performed in the following way. Typically the experimental unit is a subject and, ideally, a treatment time. The number of experimental units to be assigned to each treatment is fixed by the experimenter. From the subjects to be used in the experiment he randomly selects n_1 for the first treatment, n_2 of the remaining subjects for the second treatment, and so on. Any subject could be assigned to any of the treatments, and it is randomly determined (with sample size constraints) which treatment a subject takes. The experimental unit could, of course, be a cluster or group of subjects, along with treatment times, and could include the experimenter or technician who administers the treatment to which the unit is assigned.

The systematic procedure to be described in this chapter for determining significance of t for an independent t test by means of the randomization test procedure is one described by Fisher (1936). Fisher explained how to derive a reference set of the test statistic by repeatedly mixing and dividing two sets of sample data. The application was nonexperimental and required random sampling of two populations to test their identity. Pitman (1937) described the same computational procedure and considered its application to both experimental and nonexperimental data. He showed that the procedure of determining significance could be applied in experimental situations where "the two samples together form the whole population" (p. 129) and indicated the practical value of this area of application, where random sampling is not required. It is Pitman's rationale for the application of randomization tests to nonrandom samples that is of major importance in the typical experimental setting.

For single-subject experiments the experimental unit is a treatment time. The number of treatment times to be assigned to each treatment is fixed by the experimenter. From the treatment times to be used in the experiment the experimenter randomly selects n_1 for the first treatment, n_2 of the remaining times for the second treatment, and so on. Given this type of random assignment, the computer programs in this chapter are applicable to single-subject as well as multiple-subject data.

ANOVA and t tests are used for detecting differences between experimental treatments where it is expected that some treatments will be more effective than others. The independent t test can be regarded as a special case of one-way ANOVA: the case where there are only two treatments. However, since t tests were developed before ANOVA they tend, for historical reasons, to be used for statistical analysis although ANOVA can accommodate data from experiments where there are only two treatments as well as from those with more treatments. Since ANOVA is the more general technique, it is more useful.

The simultaneous comparison of more than two treatment groups can be advantageous to a researcher. But a significant F for a comparison of several treatments does not permit one to conclude which particular treatments differ from each other in their effects; the only justifiable statistical inference from a significant overall F is that the treatments do not all have identical effects. One-way ANOVA is not sensitive to treatment differences when many of the treatments have almost identical effects and only one or two are quite different in their effects. Whether four out of five treatments have the same effect and the fifth has a different effect cannot be determined without more specific comparisons. Follow-up tests comparing individual treatments in pairs by the use of t tests are sometimes conducted, and either one-tailed or two-tailed t tests can be used for this purpose.

In view of the t test being, in a sense, a special case of ANOVA, why are both tests considered in this chapter? One-tailed and two-tailed tests are used in conjunction with the t test, and so the determination of one-tailed as well as two-tailed significance by randomization tests can be illustrated by means of t tests. This is important because, with a randomization test, the one-tailed probability value is not necessarily half of the two-tailed probability.

4.2 ONE-WAY ANOVA WITH SYSTEMATIC DATA PERMUTATION

We will first consider a randomization test, based on systematic permutation, that can be used to determine significance of results for a comparison of two or more treatments, whether the number of subjects per treatment is equal or unequal. In the case of two treatments the P-value provided by this procedure is the same as the two-tailed value for the t test, which will be discussed later.

A systematic way of listing data permutations is necessary for the systematic data permutation method, to ensure that all data permutations are considered. We will now consider how permutations for one-way ANOVA can be listed, making use of a numerical example.

Example 4.1

We have two measurements for treatment A, two for treatment B, and three for treatment C, as follows:

A	B	C
16, 18	18, 23	25, 14, 17

To systematically list the $7!/2!2!3! = 210$ data permutations, we first assign index numbers 1 to 7 to the seven obtained measurements in the following way:

A	B	C
16, 18	18, 23	25, 14, 17
1, 2	3, 4	5, 6, 7

Now we permute the index numbers instead of the measurement values. The obtained data provide the first permutation, which is represented by the index numbers as follows:

A	B	C
1, 2	3, 4	5, 6, 7

This permutation of index numbers can be represented as a seven-digit number: 1234567. The permutations of the index numbers are listed in order of magnitude from the smallest seven-digit number, which represents the obtained results, to the largest seven-digit number, keeping the index values in ascending order within a treatment. The requirement that the only permutations listed be those where index numbers are in ascending order within treatments ensures that in the list there will be no redundancy resulting from the same combination appearing in a different order within a treatment. The following listing shows the first two and last two permutations of index numbers, when listed by the described procedure:

Permutation number	A	B	C
1	1 2	3 4	5 6 7
2	1 2	3 5	4 6 7
...	...	...	...
209	6 7	3 5	1 2 4
210	6 7	4 5	1 2 3

The general computer program that is suitable for either equal or unequal sample sizes permutes the data in the sequence given by the above listing procedure.

4.3   F FOR ONE-WAY ANOVA

Table 1 is a summary table for one-way ANOVA. In the table, k is the number of treatments and N is the total number of subjects. SS_B and SS_W are defined as:

$$SS_B = \sum\left(\frac{T_i^2}{n_i}\right) - \frac{T^2}{N} \tag{4.1}$$

$$SS_W = \Sigma X^2 - \sum\left(\frac{T_i^2}{n_i}\right) \tag{4.2}$$

where T_i and n_i refer to the total and the number of subjects, respectively, for a particular treatment, and where T and N refer to the grand total and the total number of subjects for all treatments, respectively.

4.4   AN EQUIVALENT TEST STATISTIC

It will now be shown that $\Sigma(T_i^2/n_i)$ is an equivalent test statistic to F for one-way ANOVA, for determining significance by data permutation. The formula for F in Table 1 is equivalent to $(SS_B/SS_W) \times [(N - k)/(k - 1)]$. Inasmuch as $[(N - k)/(k - 1)]$ is a constant multiplier over all data permutations, its elimination has no effect on the ordering of the data permutations with respect to the test statistic value. Thus SS_B/SS_W is an equivalent test statistic to F. The total sum of squares SS_T is a constant over all data permutations and $SS_T = SS_B + SS_W$; consequently SS_B and SS_W must vary inversely over the data permutations, which in turn implies that SS_B and SS_B/SS_W show the same direction of change over all data per-

TABLE 1   Summary Table for One-Way ANOVA

Source of variation	Sum of squares	Degrees of freedom	Mean square	F
Treatment	SS_B	$k - 1$	$SS_B/(k - 1)$	$\dfrac{SS_B/(k - 1)}{SS_W/(N - k)}$
Error	SS_W	$N - k$	$SS_W/(N - k)$	
Total	SS_T	$N - 1$		

mutations. Therefore SS_B is an equivalent test statistic to SS_B/SS_W and, of course, to F. The final step is to show that $\Sigma(T_i^2/n_i)$ is an equivalent test statistic to SS_B. The grand total T and the total number of subjects N are constants over all data permutations, and so T^2/N in formula (4.1) is a constant whose removal would change each test statistic value by the same amount, leaving the order unchanged. Thus the term $\Sigma(T_i^2/n_i)$ in formula (4.1) is an equivalent test statistic to SS_B and, therefore, to F.

For determining significance by the use of a randomization test, therefore, the computation of $\Sigma(T_i^2/n_i)$ as a test statistic will give the same P-value as the computation of F. It is easier than F to compute, and so it will be used to determine the significance of F for the general-purpose (equal-or-unequal-sample-size) computer programs for one-way ANOVA. For equal sample size ANOVA, ΣT_i^2 is an equivalent test statistic to $\Sigma(T_i^2/n_i)$ and F, and will be the test statistic used for determining significance.

Instead of using the notation $\Sigma(T_i^2/n_i)$, hereafter we will omit the subscripts and use the simplified notation $\Sigma(T^2/n)$. Similarly, instead of using ΣT_i^2, we will use ΣT^2 to refer to the sum of the squared totals for the individual treatments.

We will now consider the principal operations to be preformed by a computer in finding the significance of F for one-way ANOVA by systematic data permutation:

Step 1. Arrange the data in tabular form (grouping the measurements according to treatment) and assign index numbers 1 to N to the measurements, beginning with treatment A. For example, with two subjects for treatment A, three for treatment B, and four for treatment C, index numbers 1 and 2 are assigned to the two A measurements, 3, 4, and 5 to the B measurements, and 6, 7, 8, and 9 to the C measurements.

Step 2. Compute GT, the grand total of all of the measurements.

Step 3. Generate a systematic sequence of permutations of the index numbers associated with the measurements, so that there are n_A index numbers for A, n_B for B, and so on. For each of these permutations, perform steps 4 to 6 on the measurements associated with the index numbers. The first permutation represents the obtained results, and so steps 4 to 6 for the measurements associated with the first permutation provide the obtained test statistic value. There will be $N!/n_A!n_B! \cdots n_k!$ permutations of the N measurements for the k groups.

Step 4. Compute $T_A, \ldots, T_{k-1}$, the totals for all of the treatments but the last.

Step 5. Compute $T_k = GT - (T_A + \cdots + T_{k-1})$, the total for the last treatment.

Step 6. Add $T_A{}^2/n_A, \ldots, T_k{}^2/n_k$ to get $\Sigma(T^2/n)$, the test statistic.

Step 7. Compute the probability for F as the proportion of the $N!/n_A! n_B! \cdots n_k!$ permutations (including the first permutation) that provide as large a value of $\Sigma(T^2/n)$ as the obtained test statistic value, the value computed for the first permutation.

Example 4.2

An experimenter randomly assigns nine subjects to three treatments, with two subjects to take treatment A, three to take B, and four to take C. Table 2 shows the results. The grand total of the measurements is 107. For the first permutation, given in Table 2, $T_A = 14$, $T_B = 29$, and $T_C = 107 - (14 + 29) = 64$. The test statistic value for the obtained results (that is, the obtained test statistic value) is $(14)^2/2 + (29)^2/3 + (64)^2/4 = 1,402\text{-}1/3$. The second permutation (not shown) is the one where treatment A has the same measurements, but treatment B contains the measurements for index numbers 3, 4, and 6, and treatment C contains the measurements for index numbers 5, 7, 8, and 9. For the second permutation, the test statistic value is $(14)^2/2 + (37)^2/3 + (56)^2/4 = 1,338\text{-}1/3$. There are $9!/2! 3! 4! = 1,260$ permutations for which test statistic values are computed. The obtained test statistic value is larger than any of the other 1,259 test statistic values, and so the probability for the observed results is $1/1,260$, or about 0.0008.

TABLE 2   Experimental Results

Treatment A		Treatment B		Treatment C	
Index number	Measurement	Index number	Measurement	Index number	Measurement
1	6	3	9	6	17
2	8	4	11	7	15
		5	9	8	16
				9	16

Program 4.1 goes through the preceding seven steps to determine the significance of F for one-way ANOVA by means of a randomization test based on systematic data permutation. Program 4.1 can be tested by using the following data, for which the significance value is 2/1,260 or about 0.0016:

A: 1, 2
B: 3, 4, 5
C: 7, 8, 9, 10

4.5   ONE-WAY ANOVA WITH EQUAL SAMPLE SIZES

The test statistic in Program 4.1 for obtaining the significance of F is $\Sigma(T^2/n)$. When the sample sizes are equal, this test statistic is equal to $(\Sigma T^2)/n$, where n is the number of subjects per sample. For situations with equal sample sizes, n is a constant divisor over all data permutations, making ΣT^2 an equivalent test statistic to $(\Sigma T^2)/n$ and thus equivalent to F. Use of the simpler test statistic simplifies the program and reduces computational time slightly.

A short cut that saves considerable time when the sample sizes are equal is to use a subset of the entire set of data permutations. In Section 3.3 it was explained that to determine significance by systematic data permutation for a nondirectional test statistic like F or $|t|$, it is necessary to compute only 1/k! of the total number of data permutations for k treatments with equal sample sizes. That is because for each division of the data into k groups there are k! ways of assigning the k treatment designations to those groups, all of which must give the same test statistic value. We will now consider a permutation listing procedure for generating a subset of data permutations that will give the same P-value as the set of all data permutations when the sample sizes are equal.

Example 4.3

The value of ΣT^2 for a difference between three groups of equal size is the same for the data permutation associated with the division of index numbers into A: 3, 7, 8; B: 2, 5, 9; C: 1, 4, 6, as for any case where the measurements are separated into three groups in the same way but with transposition of the group labels. For example, the data permutation associated with the division of index numbers into A: 2, 5, 9; B: 1, 4,

PROGRAM 4.1 One-way ANOVA:
Systematic Permutation

```
        ((INPUT))

    OBTAIN=NGE=TOTAL=NN=NPERM=W=0
    READ (5,---)NGRPS, (N(I),I=1,NGRPS)
    DO 2 I=1,NGRPS
    K=NN+1
    NN=NN+N(I)
    IF(I.EQ.NGRPS)W=1
    DO 1 J=K,NN
    READ(5,---)DATA(J)
    TOTAL=TOTAL+DATA(J)
    INDEX(J)=J
1   WOR(J)=W
2   N(I)=NN
    IG=NGRPS-1
3   TEST=SUBTOT=K=0
    NPERM=NPERM+1
    DO 5 I=1,IG
    J=K+1
    K=N(I)
    SUM=0
    DO 4 L=J,K
4   SUM=SUM+DATA(INDEX(L))
    TEST=TEST+SUM**2/(K-J+1)
5   SUBTOT=SUBTOT+SUM
    TEST=TEST+(TOTAL-SUBTOT)**2/(N(NGRPS)-K)
    IF(NPERM.EQ.1)OBTAIN=TEST
    IF(TEST.GE.OBTAIN)NGE=NGE+1
    J=IG
    I=N(IG)
6   WOR(INDEX(I))=1
7   IF(INDEX(I).EQ.NN)GO TO 9
    INDEX(I)=INDEX(I)+1
8   IF(WOR(INDEX(I)).EQ.0)GO TO 7
    WOR(INDEX(I))=0
    IF(I.EQ.N(IG))GO TO 3
    I=I+1
    INDEX(I)=INDEX(I-1)
    IF(I.NE.N(J)+1)GO TO 7
    INDEX(I)=1
    J=J+1
    GO TO 8
9   I=I-1
    IF(J.NE.1.AND.I.EQ.N(J-1))J=J-1
    IF(I.NE.0)GO TO 6
    PROB=FLOAT(NGE)/NPERM

        ((OUTPUT))
```

TABLE 3   List of Index
Number Permutations

A	B	C
1 2 3	4 5 6	7 8 9
1 2 3	4 5 7	6 8 9
. . . . .	. . . . .	. . . . .
* 1 4 6	2 5 9	3 7 8
. . . . .	. . . . .	. . . . .
* 1 4 6	3 7 8	2 5 9
. . . . .	. . . . .	. . . . .
2 5 9	1 4 3	6 7 8
* 2 5 9	1 4 6	3 7 8
. . . . .	. . . . .	. . . . .
* 2 5 9	3 7 8	1 4 6
2 5 9	4 6 7	1 3 8
. . . . .	. . . . .	. . . . .
* 3 7 8	1 4 6	2 5 9
. . . . .	. . . . .	. . . . .
* 3 7 8	2 5 9	1 4 6
. . . . .	. . . . .	. . . . .
7 8 9	4 5 6	1 2 3

6; C: 3, 7, 8, would give the same value of ΣT^2. Inasmuch as ΣT^2 is not affected by the direction of difference between means, clearly the value is the same for one permutation as for the other. In fact, for the three sets of measurements, there are 3! = 6 different ways of assigning the three group designations A, B, and C, each of which gives the same value of ΣT^2 as the others. The systematic listing procedure, described in Section 4.2, for listing all possible data permutations would provide a list that includes all six of these. Table 3 shows part of the list of index number permutations. The asterisks indicate the permutations that involve the separations of index numbers into (1, 4, 6), (2, 5, 9), and (3, 7, 8). A different listing procedure will now be described which will require listing only one-sixth of all permutations when the three groups have equal sample sizes to get a sampling distribution of ΣT^2 that is equivalent, for computing probabilities, to the sampling distribution based on all data permutations. The distribution of one-sixth of the assignments will involve no redundancy of the kind where there is the same division of measurements with transposition of group labels.

The listing procedure to eliminate redundancy, like the procedure in Section 4.2, arranges index numbers in ascending order within groups for each permutation, and the permutations are listed sequentially in ascending order of magnitude of the index numbers where the N index numbers for a permutation are regarded as an N-digit number. To provide the restriction required to eliminate the redundancy, only those permutations are listed where the smallest index numbers for treatments are in ascending order from A to C. Thus, of the six permutations with asterisks, only one would be listed, namely, A: 1, 4, 6; B: 2, 5, 9; C: 3, 7, 8. That is the only one of the six permutations where the three-digit numbers formed by the index numbers are in ascending order from A to C. Such a restriction permits every partition of measurements to be listed once and only once; relabeling the group designations for a particular partition is not permitted. This, then, provides the one-sixth of the permutations that contain all partitions of the measurements when group designations are ignored; therefore duplication of ΣT^2 values resulting from the same partition being simply relabeled is eliminated. (The obtained results, with index numbers of A: 1, 2, 3; B: 4, 5, 6; C: 7, 8, 9, give the first data permutation in the nonredundant list.)

To generalize, the above listing procedure permits the significance of F to be determined by use of a randomization test employing only $1/k!$ of all possible data permutations when all k treatments have equal sample sizes.

The following steps are used to determine significance of F by systematic permutation when only the nonredundant permutations are considered:

Step 1. Arrange the data in tabular form, grouping the measurements according to treatment. Assign index numbers 1 to N to the measurements, beginning with treatment A.

PROGRAM 4.2 One-way ANOVA—Equal N: Systematic Permutation

```
            ((INPUT))

    READ (5,---)NGRPS,N
    OBTAIN=TOTAL=W=NN=O
    DO 2 I=1,NGRPS
    IF(I.EQ.NGRPS)W=1
    M=NN+1
    NN=NN+N
    SUM=0
    DO 1 J=M,NN
    READ(5,---)DATA(J)
    SUM=SUM+DATA(J)
    INDEX(J)=J
  1 WOR(J)=W
    OBTAIN=OBTAIN+SUM**2
  2 TOTAL=TOTAL+SUM
    NGE=NPERM=1
    IG=NGRPS-1
    MM=(NGRPS-1)*N
  3 I=MM
  4 WOR(INDEX(I))=1
  5 IF(INDEX(I).EQ.NN)GO TO 9
    INDEX(I)=INDEX(I)+1
    IF(WOR(INDEX(I)).EQ.O)GO TO 5
    WOR(INDEX(I))=0
    IF(I.EQ.MM)GO TO 6
    I=I+1
    INDEX(I)=INDEX(I-1)
    IF(I/N*N+1.EQ.I)INDEX(I)=0
    GO TO 5
  6 TEST=SUBTOT=K=0
    NPERM=NPERM+1
    DO 8 I=1,IG
    J=K+1
    K=K+N
    SUM=0
    DO 7 L=J,K
  7 SUM=SUM+DATA(INDEX(L))
    TEST=TEST+SUM**2
  8 SUBTOT=SUBTOT+SUM
    TEST=TEST+(TOTAL-SUBTOT)**2
    IF(TEST.GE.OBTAIN)NGE=NGE+1
    GO TO 3
  9 I=I-1
    IF(I/N*N+1.NE.I)GO TO 4
    WOR(INDEX(I))=1
    I=I-1
    IF(I.NE.O)GO TO 4
    PROB=FLOAT(NGE)/NPERM

            ((OUTPUT))
```

Step 2.   Compute GT, the grand total of all the measurements.
Step 3.   Use the redundancy-eliminating listing procedure to generate a systematic sequence of permutations. For each of these permutations, including the first (which represents the observed results), perform steps 4 to 6.
Step 4.   Compute $T_A, \ldots, T_{k-1}$, that is, the totals for each of the treatments except the last.
Step 5.   Compute $T_k = GT - (T_A + \cdots + T_{k-1})$, the total for the last treatment.
Step 6.   Add $T_A{}^2, \ldots, T_k{}^2$ to get ΣT^2, the test statistic.
Step 7.   Compute the probability for F as the proportion of the permutations providing as large a value of ΣT^2 as the obtained test statistic value (the value for the first, or obtained, permutation).

Program 4.2 goes through these steps in determining the significance of F by use of a randomization test employing non-redundant data permutation. Program 4.2 can be tested by employing it with the following data, for which the P-value is 18/1,680, or about 0.0107:

A: 1, 2, 4
B: 3, 5, 6
C: 7, 8, 9

4.6   ONE-WAY ANOVA WITH RANDOM DATA PERMUTATION

Reducing the number of permutations and test statistic computations to be performed through the use of a listing procedure that omits redundant permutations (ones that are "mirror images" of permutations already listed) makes the use of a randomization test to determine the significance of F for one-way ANOVA more practical than otherwise, but there are two disadvantages. In the first place, such a procedure can be employed only when the sample sizes are equal. Secondly, although performing only 1/k! of the permutations is a considerable reduction in the amount of work required, there are still many permutations to be generated if the total number of permutations runs into millions, billions, or trillions. Thus, to make the determination of significance by permutation practical for relatively large samples, an alternative procedure is required.

An alternative method of permuting that is highly practical, even for large sample sizes, is the random permutation method, a method wherein a fixed number of permutations are randomly selected and the probability or significance value is based upon those permutations. Although it is most practical to use random permutation when the number of possible permutations is large, it may also be useful when the total number of possible permutations is only a few thousand.

The random permutation procedure incorporated into the programs for one-way ANOVA and the independent t test in this chapter is an analog of the procedure of random selection described by Green (1963, pp. 172–173).

There is no need for a separate random permutation method for the special case where the sample sizes are equal. Thus the test statistic to be employed for the random permutation method is that of the systematic data permutation method which is appropriate for both equal and unequal samples: $\Sigma(T^2/n)$.

The following steps show the computational procedure:

Step 1. Arrange the data in the form of a table, grouping the measurements according to treatment, and assign index numbers 1 to N to the measurements in sequence, beginning with treatment A.

Step 2. Compute GT, the grand total of all the measurements.

Step 3. Go through steps 5 to 7 for the observed data to compute the obtained test statistic value.

Step 4. Use a random number generation algorithm that will select n_A index numbers without replacement from the N index numbers to assign the corresponding measurements to treatment A. From the remaining index numbers randomly select without replacement n_B index numbers to assign the corresponding measurements to treatment B. Continue assigning until all of the index numbers have been assigned to treatments. Each assignment of index numbers to every treatment constitutes a permutation of the data. Where NPERM is the requested number of permutations, perform NPERM − 1 permutations and for each permutation go through steps 5 to 7.

Step 5. Compute $T_A, \ldots, T_{k-1}$, the totals of each of the treatments except the last.

Step 6. Compute $T_k = GT - (T_A + \cdots + T_{k-1})$, the total for the last treatment.

Step 7. Add $T_A^2/n_A, \ldots, T_k^2/n_k$ to get $\Sigma(T^2/n)$, the test statistic value.

PROGRAM 4.3   One-way ANOVA:
Random Permutation

```
            ((INPUT))
    READ(5,---)NGRPS,(N(I),I=1,NGRPS)
    TOTAL=OBTAIN=NN=0
    READ(5,---)NPERM
    DO 2 I=1,NGRPS
    MM=NN+1
    NN=NN+N(I)
    SUM=0
    DO 1 J=MM,NN
    READ(5,---)DATA(J)
1   SUM=SUM+DATA(J)
    TOTAL=TOTAL+SUM
2   OBTAIN=OBTAIN+SUM**2/N(I)
    NGE=1
    IG=NGRPS-1
    DO 5 I=2,NPERM
    TEST=SUBTOT=MM=0
    DO 4 J=1,IG
    L=MM+1
    MM=MM+N(J)
    SUM=0
    DO 3 K=L,MM
    KK=K+RANF(0)*(NN-K+1)
    X=DATA(KK)
    DATA(KK)=DATA(K)
    DATA(K)=X
3   SUM=SUM+X
    SUBTOT=SUBTOT+SUM
4   TEST=TEST+SUM**2/N(J)
    TEST=TEST+(TOTAL-SUBTOT)**2/N(NGRPS)
5   IF(TEST.GE.OBTAIN)NGE=NGE+1
    PROB=FLOAT(NGE)/NPERM
            ((OUTPUT))
```

Step 8. Compute the probability for F as the proportion of the
    NPERM permutations (the NPERM − 1 performed by the
    computer plus the obtained data configuration) that provide
    a test statistic value as large as the obtained test statistic
    value.

Program 4.3 goes through these eight steps for a randomization test to determine the significance of F by random permutation. Program 4.3 can be tested with the following data, for

which the significance value by systematic data permutation is
6/1,260, or about 0.0048:

A: 1, 2
B: 3, 4, 5
C: 6, 7, 8, 9

4.7   ANALYSIS OF COVARIANCE

Analysis of covariance is a form of analysis of variance with
statistical control over the effect of extraneous variables. It
consists of an analysis of variance performed on transformed
dependent variable measures that express the magnitude of the
dependent variable relative to the value that would be predicted
from a regression of the dependent variable on the extraneous
variable. For instance, consider a comparison of the effects of
two drugs on body weight. An analysis of variance performed
to test the difference between weights of the subjects in the
two treatment groups might fail to give significant results simply
because the large variability of the pre-experimental body
weight made the statistical test relatively insensitive to the difference in the drug effects. One way to control for the effect
of pre-experimental body weight is by applying analysis of variance to difference scores: that is, to the gain in weight.
Generally speaking, applying analysis of variance to difference
scores is regarded as less suitable than analysis of covariance
for such an application. Analysis of covariance, unlike the
application of analysis of variance to difference scores, takes
into consideration the degree of correlation between the control
or concomitant variable (such as pre-experimental body weight)
and the dependent variable (such as final body weight), thereby
tending to provide a more powerful test than would result from
analysis of the difference scores.
     Analysis of covariance also can control for the effects of
an extraneous variable in situations where there are no relevant
difference scores, as when the concomitant variable is different
in kind from the dependent variable. For example, one may
want to control for differences in intelligence in a comparison of
various methods of learning. Subjects may differ considerably
in intelligence within and between groups, and insofar as variation in intelligence results in variation in learning performance
that is not a function of the experimental treatments, it is desirable to remove this effect. Analysis of covariance can be
used to increase the power of the test by controlling for differences in intelligence. Analysis of covariance would compare

learning performance under the alternative methods, where the learning performance of a person is expressed relative to the performance expected of a person of that level of intelligence, where the "expected" performance is that determined by the regression of performance on intelligence.

Analysis of covariance involves assumptions in addition to those of analysis of variance, such as linearity of regression and a common slope of regression lines over various treatment groups, so that even if users were not concerned about non-randomness of sampling, there might be a need for a nonparametric procedure. Let us consider, then, how analysis of co-variance can be performed with significance determined by the randomization test procedure. First, before conducting the experiment, a measure of the concomitant variable (e.g., intelligence) is taken. (By taking the measurement at that time, it is ensured that the concomitant variable is not affected by the treatment.) Then subjects are assigned randomly to the alternative treatments, and measures of the dependent variable are taken. An analysis of covariance then is carried out in the usual manner to provide the test statistic F that reflects the difference between treatments when there is control over the effect of the concomitant variable. The data are then permuted and for each data permutation the test statistic F is computed, to provide the reference set of data permutations to which the obtained F is referred for determination of significance.

4.8   INDEPENDENT t TEST WITH SYSTEMATIC DATA PERMUTATION

The equivalence of F and $|t|$ as randomization test statistics permits ANOVA procedures to be employed for determining the significance for a two-tailed t test. Program 4.1 can be used to determine the P-value for a two-tailed t test, whether the two sample sizes are equal or unequal; but Program 4.2 is slightly more efficient for a two-tailed t test when the sample sizes are equal, because only half of the permutations need be used.

For one-tailed t tests, however, the ANOVA programs are not particularly useful. The one-tailed probability values are not given directly by the programs, and they cannot always be derived from the two-tailed probabilities that are given. For probability combining, a technique discussed in Chapter 6 that can be quite useful, it is essential that there be a P-value associated with a one-tailed t test, even when the direction of difference is *incorrectly* predicted. Let us now consider what

such a P-value means. It is conventional in computing a value
of t by formula (1.1), where $\bar{X}_A - \bar{X}_B$ is the numerator, to
designate as treatment A the treatment predicted to provide the
larger mean. In that way a positive value of t will be obtained
if the direction of difference between means is correctly predicted and a negative value of t will be obtained if the obtained
direction is opposite to the predicted one. Thus a measure of
the extent to which a value of t is consistent with the predicted
direction of effect is its magnitude, taking into consideration
the sign associated with the value of t. Therefore we define
the one-tailed probability for t for a randomization test as the
probability, under H_0, of getting such a large value of t as the
obtained value. If the direction of difference between means is
correctly predicted, the obtained t will have a positive value
and the proportion of data permutations with such a large t may
be fairly small; but if the direction of difference is incorrectly
predicted, the t value will be negative and the probability will
tend to be large.

When the sample sizes are equal, the distribution of t under
systematic data permutation is symmetrical about 0, there being
associated with every positive t value a negative t with the same
absolute value. If an absolute value of t of 2.35 has a P-value
of 0.10 by data permutation, then 5% of the t's are greater than
or equal to +2.35, and 5% are less than or equal to −2.35. If
we have correctly predicted the direction of difference between
means (thereby obtaining a positive value of t) the one-tailed
probability will be 0.05, half of the two-tailed value. But
suppose we predicted the wrong direction. What is the one-
tailed probability in that case? In other words, what proportion
of the data permutations give a t value greater than, or equal
to, −2.35? We know that 95% are greater than −2.35, but we
do not know what percentage are greater than or equal to
−2.35, because we do not know how many t's equal to −2.35 are
in the distribution given by data permutation. Consequently,
although we can halve the two-tailed probability value given by
Program 4.2 with equal sample sizes and determine the one-
tailed value for a correct prediction of the direction of difference
between means, we cannot determine from the two-tailed probability the one-tailed probability for an incorrectly predicted
direction of difference.

When sample sizes are unequal, however, not even the one-
tailed probability for a *correct* prediction of the direction of
difference between means can be determined by halving the two-
tailed probability. The proportion of test statistics with a value
of $|t|$ as large as the obtained $|t|$ and with the same direction
of difference between means as the obtained direction is not

necessarily half of the proportion of $|t|$'s that are as large as the obtained $|t|$.

Example 4.4

Consider the three permutations of the measurements 2, 3, and 5 for treatments A and B, where A has one measurement and B has two:

A	B	A	B	A	B
2	3	3	2	5	2
	5		5		3
$|t| = 1.33$		0.04		8.33	

If the third permutation was the obtained permutation, the computed probability value would be 1/3 for the two-tailed t test. If we were to halve that value because we predicted that the A mean would be larger than the B mean, we would obtain a P-value of 1/6, when in fact the smallest *possible* P-value for any test statistic is 1/3 because only three assignments are possible.

For the above situation the two-tailed probability would be the same as the one-tailed probability for a correctly predicted direction of difference between means. With unequal sample sizes we cannot obtain a one-tailed probability from the two-tailed probability but must compute it separately.

The following computer programs for the independent t test will accommodate either equal or unequal sample sizes. For a two-tailed test the test statistic used is $\Sigma(T^2/n)$, which is equivalent to $|t|$. For a one-tailed test, the test statistic used is T_L (the total of the measurements for the treatment predicted to give the larger mean), which is equivalent to t, with the sign considered, as a test statistic.

We will now consider the steps to be performed by a computer in determining both two-tailed and one-tailed P-values for unequal or equal sample sizes.

Step 1. Arrange the research data in the form of a table, grouping the measurements according to treatment, where the first treatment (called treatment L) is the treatment predicted to have the larger mean, and the second treatment (called treatment S) is the treatment predicted to have the smaller mean. Assign index numbers 1 to N to the measurements, beginning with the measurements for treatment L.

Step 2. Compute GT, the grand total of all the measurements.

Step 3. Perform step 6 for the obtained data to compute the obtained one-tailed test statistic value.

Step 4. Perform steps 6 to 8 for the obtained data to compute the obtained two-tailed test statistic value.

Step 5. Generate a systematic sequence of permutations of the index numbers associated with the measurements, with n_L index numbers for treatment L and n_S for treatment S. For each of these permutations perform steps 6 to 8 on the measurements associated with the index numbers.

Step 6. Compute T_L, the total of the measurements for treatment L, to get the one-tailed test statistic value.

Step 7. Compute T_S, the total of the measurements for treatment S by subtracting T_L from GT.

Step 8. Add T_L^2/n_L and T_S^2/n_S to get $\Sigma(T^2/n)$, the two-tailed test statistic value.

Step 9. Obtain the two-tailed P-value by determining the proportion of the permutations that provide as large a two-tailed test statistic value as the obtained two-tailed value.

Step 10. Obtain the one-tailed P-value by determining the proportion of the permutations that provide a one-tailed test statistic value as large as the obtained one-tailed value.

Example 4.5

We have two measurements for treatment L and three measurements for treatment S, as shown in the following table:

Treatment L		Treatment S	
Index	Measurement	Index	Measurement
1	6	3	3
2	9	4	6
		5	9

The grand total of the measurements is 33, the mean of the L measurements is 7.5 and the mean of the S measurements is 6. There are $5!/2!3! = 10$ permutations of the data to be performed.

TABLE 4 Test Statistic Values

L index numbers	1, 2	1, 3	1, 4	1, 5	2, 3	2, 4	2, 5	3, 4	3, 5	4, 5
L measurements	6, 9	6, 3	6, 6	6, 9	9, 3	9, 6	9, 9	3, 6	3, 9	6, 9
T_L	15	9	12	15	12	15	18	9	12	15
T_S	18	24	21	18	21	18	15	24	21	18
$T_L^2/2$	112.5	40.5	72	112.5	72	112.5	162	40.5	72	112.5
$T_S^2/3$	108	192	147	108	147	108	75	192	147	108
$\Sigma(T^2/n)$	220.5	232.5	219	220.5	219	220.5	237	232.5	219	220.5

The one-tailed and two-tailed test statistic values for each of the permutations will now be computed. The first row of numbers in Table 4 indicates the 10 possible pairs of *index numbers* for treatment L, whereas the second row indicates the pairs of *measurements* associated with those index numbers. The first column represents the observed data, the experimental results. The one-tailed probability is the proportion of the T_L values that are as large as the obtained value, 15, which is 0.50. The two-tailed probability is the proportion of the $\Sigma(T^2/n)$ values that are as large as the obtained value, 220.5, which is 0.70.

Program 4.4 can be used to obtain both two-tailed and one-tailed P-values for t for equal or unequal sample sizes by systematic permutation. Program 4.4 can be tested with the following data, for which the one-tailed P-value is 1/35, or about 0.0286 and the two-tailed P-value is 2/35, or about 0.0571:

L: 4, 5, 6, 7
S: 1, 2, 3

4.9   INDEPENDENT t TEST WITH RANDOM
DATA PERMUTATION

The number of permutations for systematic data permutation does not increase as rapidly with an increase in number of subjects per group for the t test as for ANOVA where there are several groups, but the increase is rapid enough to restrict the utility of the systematic permutation method to relatively small sample sizes. With only 10 subjects for each of the two treatments the number of permutations is 184,756, and with 12 subjects per treatment the number of permutations is 2,704,156. Thus, to determine significance by means of a randomization test the random permutation method is necessary for the t test, even when samples are not very large.

For determining the probability for a two-tailed t test by random permutation, Program 4.3 can be used since F and $|t|$ are equivalent test statistics. When the groups are equal in size, the probability value given by Program 4.3 can be divided by two to give the one-tailed value where the direction of difference has been *correctly* predicted. The justification for halving the two-tailed probability value to get the one-tailed value under random permutation with equal sample sizes is not the same as that of halving the two-tailed value when there is systematic permutation. With systematic permutation, halving the two-tailed P-value is a simple way of getting exactly the same P-value as would have been obtained by directly determining the proportion

PROGRAM 4.4 Independent t
Test: Systematic Permutation

```
          ((INPUT))

     READ(5,---)N1,N2
     NN=N1+N2
     TOTAL=0
     DO 1 I=1,NN
     READ(5,---)DATA(I)
     TOTAL=TOTAL+DATA(I)
     IF(I.EQ.N1)SUM=TOTAL
1    INDEX(I)=I
     OBTONE=SUM
     OBTTWO=SUM**2/N1+(TOTAL-SUM)**2/N2
     NPERM=NGEONE=NGETWO=1
2    I=N1
3    IF(INDEX(I).EQ.NN)GO TO 6
     INDEX(I)=INDEX(I)+1
     IF(I.EQ.N1)GO TO 4
     I=I+1
     INDEX(I)=INDEX(I-1)
     GO TO 3
4    NPERM=NPERM+1
     SUM=0
     DO 5 I=1,N1
5    SUM=SUM+DATA(INDEX(I))
     TESTTWO=SUM**2/N1+(TOTAL-SUM)**2/N2
     IF(TESTTWO.GE.OBTTWO)NGETWO=NGETWO+1
     TESTONE=SUM
     IF(TESTONE.GE.OBTONE)NGEONE=NGEONE+1
     GO TO 2
6    I=I-1
     IF(I.NE.0)GO TO 3
     PROBONE=FLOAT(NGEONE)/NPERM
     PROBTWO=FLOAT(NGETWO)/NPERM

          ((OUTPUT))
```

---

of the permutations providing as large a value of ΣT^2 as the obtained value, with the difference between means in the same direction as for the obtained data. But with the random permutation method, in the sampling distribution of, say, 1,000 permutations, there is no assurance that half of the permutations giving a test statistic as large as the obtained value will have a difference between means in one direction and half in the other. Consequently, the halving of the two-tailed P-value does not necessarily provide the proportion of the 1,000 permutations that would give a test statistic as large as the obtained value with the obtained direction of difference between means.

The validity of halving the two-tailed probability for random permutation tests (with equal sample sizes) can be demonstrated through consideration of the interpretation that is to be placed on the one-tailed P-value. We want the procedure for determining the one-tailed probability values to be such that, if the null hypothesis of identical treatment effects is true, the probability of getting a P-value as small as P is no greater than P. For example, in the long run no more than 5% of the one-tailed P-values should be as small as 0.05. The validity of the *two-tailed* random permutation probability is shown in Section 3.4, which demonstrates that the probability associated with any test statistic value can be determined in the manner in which it was determined for ΣT^2 for the random permutation method. For example, we can conclude that when H_0 is true, the probability of getting a two-tailed probability as small as 0.10 is no greater than 0.10. With equal sample sizes, half of the differences between means will be in one direction and half in the other. Since the random sampling distribution of permutations is a random sample from the population of all permutations, for any difference between means the probability that it is in the predicted direction is 1/2. Thus, for cases where a two-tailed probability is as small as 0.10, the probability is 1/2 that the difference is in the predicted direction when H_0 is true. Consequently, under H_0 the probability of getting a two-tailed P-value by random permutation as small as 0.10 with the predicted direction of difference is 1/2 of 0.10, or 0.05. This, then, justifies the determination of a one-tailed probability value by random permutation by means of dividing the two-tailed probability by 2, in the cases where the sample sizes are equal.

Program 4.3, which is for one-way ANOVA with significance determined by random data permutation, can be used to provide P-values for the independent t test, whether the sample sizes are equal or unequal. If the sample sizes are equal, halving the P-value given by the program gives the one-tailed value for the t test, if the direction of difference has been correctly predicted.

But for unequal sample sizes, with random permutation as with systematic permutation, it is necessary to determine the one-tailed probability value for a correctly predicted direction of difference directly rather than by halving the two-tailed P-value. The test statistic for the two-tailed P-value is $\Sigma(T^2/n)$ and for the one-tailed P-value is T_L, the total of the measurements for the treatment predicted to have the larger mean.

For the reasons given in Section 4.8 for systematic data permutation, one-tailed P-values where the direction of difference was *incorrectly* predicted (useful P-values for probability combining) cannot be derived from the two-tailed P-values given by Program 4.3 for either equal or unequal sample sizes but must be directly computed.

The following steps are required to determine one-tailed probabilities for t by random permutation, for equal or unequal sample sizes:

Step 1. Arrange the research data in the form of a table, where the measurements are grouped by treatment, where the first treatment, called treatment L, is the treatment predicted to have the larger mean and the second treatment, treatment S, is the treatment predicted to have the smaller mean. Assign index numbers 1 to N to the measurements, beginning with the measurements for treatment L.

Step 2. Compute GT, the grand total of all of the measurements.

Step 3. Perform step 6 for the obtained data to compute the obtained one-tailed test statistic value.

Step 4. Perform steps 6 to 8 for the obtained data to compute the obtained two-tailed test statistic value.

Step 5. Use a random number generation algorithm that will select N_L index numbers without replacement from the N index numbers to assign the associated N_L measurements to treatment L. The remaining measurements are for treatment S. Each assignment of N_L measurements to treatment L constitutes a single permutation of the data. Perform NPERM $-$ 1 permutations and for each permutation go through steps 6 to 8.

Step 6. Compute T_L, the total of the measurements for treatment L, to get the one-tailed test statistic value.

Step 7. Compute T_S, the total of the measurements for treatment S, by subtracting T_L from GT.

Step 8. Add T_L^2/n_L and T_S^2/n_S to get $\Sigma(T^2/n)$, the two-tailed test statistic value.

Step 9. Obtain the two-tailed probability by determining the proportion of the permutations that provide as large a two-tailed test statistic value as the obtained two-tailed value, given by step 4.

Step 10. Obtain the one-tailed probability by determining the proportion of the NPERM permutations that provide as large a one-tailed test statistic value as the obtained one-tailed value computed in step 3.

Program 4.5 performs these operations. It can be tested with the following data, for which the one-tailed probability is 1/35, or about 0.0286, and the two-tailed probability is 2/35, or about 0.0571:

PROGRAM 4.5
Independent t Test:
Random Permutation

```
        ((INPUT))

  READ(5,---)N1,N2
  NN=N1+N2
  TOTAL=0
  READ(5,---)NPERM
  DO 1 I=1,NN
  READ(5,---)DATA(I)
  TOTAL=TOTAL+DATA(I)
1 IF(I.EQ.N1)SUM=TOTAL
  OBTONE=SUM
  OBTTWO=SUM**2/N1+(TOTAL-SUM)**2/N2
  NGEONE=NGETWO=1
  DO 3 I=2,NPERM
  SUM=0
  DO 2 J=1,N1
  K=J+RANF(0)*(NN-J+1)
  X=DATA(K)
  DATA(K)=DATA(J)
  DATA(J)=X
2 SUM=SUM+X
  TESTTWO=SUM**2/N1+(TOTAL-SUM)**2/N2
  IF(TESTTWO.GE.OBTTWO)NGETWO=NGETWO+1
  TESTONE=SUM
3 IF(TESTONE.GE.OBTONE)NGEONE=NGEONE+1
  PROBONE=FLOAT(NGEONE)/NPERM
  PROBTWO=FLOAT(NGETWO)/NPERM

        ((OUTPUT))
```

L: 4, 5, 6, 7
S: 1, 2, 3

4.10   INDEPENDENT t TEST AND PLANNED COMPARISONS

Where k is the number of treatments, there are $(k)(k - 1)/2$ pairs of treatments which could be subjected to a t test. For example, for five treatments there are $(5)(4)/2 = 10$ pairs of treatments. If an experiment is conducted for the purpose of applying t tests to a small number of specified pairs of treatments rather than to all possible pairs, the comparisons that are made are called *planned comparisons*.

When an experiment is performed to provide the basis for planned comparisons, the computation of F and the determination of its significance are unnecessary. Let us see how a t test for a planned comparison is performed when significance is determined by a randomization test.

Suppose we have randomly assigned subjects to each of five treatments, A, B, C, D, and E. One of the planned comparisons is of treatments A and C, to see if they have the same effect. The test is performed as if A and C were the only treatments in the experiment. The A and C measurements are repeatedly divided between A and C to determine the significance of t by data permutation. Thus, Program 4.4 or 4.5 could be employed to give the one-tailed and two-tailed probabilities for the test. Other planned comparisons within the five treatments would be conducted in the same way.

4.11 INDEPENDENT t TEST AND MULTIPLE COMPARISONS

When comparisons are not planned before an experiment but, instead, t tests are applied to all (or virtually all) possible pairs of treatments, such comparisons are called post hoc comparisons. In the belief that these comparisons should be treated differently from planned comparisons, special "multiple comparisons" procedures were developed for post hoc comparisons.

Before special multiple comparisons procedures were developed, it was common to determine significance for post hoc t tests in the same way as for planned t tests, but it was generally assumed that t tests would be conducted only if the overall F was significant at the 0.05 level. This means of determining significance for post hoc t tests is still used by some experimenters.

A number of alternative procedures have been proposed for determining significance for multiple comparisons. Miller (1966) presents a large number of parametric and nonparametric techniques. Experimental design books (e.g., Keppel, 1973, pp. 133–163; Myers, 1966, pp. 328–337; Winer, 1971, pp. 185–204) frequently present several parametric procedures for the reader to consider. Many of the commonly used procedures require reference to a table of the *studentized range statistic*, which, like the t table, is based on assumptions of random sampling, normality, and homogeneity of variance. The validity of those procedures thus depends on the tenability of the assumptions underlying the table of the studentized range statistic. Therefore, it is recommended that a distribution-free multiple comparisons procedure be employed. *Fisher's modified least significant difference* (modified LSD) is such a procedure, since it does not require reference to the studentized range statistic table or any other table.

Winer (1971, pp. 199—200) discusses Fisher's modified LSD procedure. We will now consider how to apply that procedure when we determine P-values by data permutation. To use the modified LSD procedure, we set an *experiment-wise error rate*, or α. The experiment-wise error rate is the probability of falsely rejecting at least one individual-comparison H_0 in an experiment.

Example 4.6

An experimenter sets an experiment-wise error rate of 0.05 for multiple comparisons of four treatments. For four treatments there are six possible treatment pairs. The P-value is determined for each of the six comparisons by use of the randomization test procedure, where for each comparison only the data for the compared treatments are permuted. Then each P-value is multiplied by 6 (the number of comparisons) to give the adjusted P-value. If the adjusted value is as small as 0.05, the t value for that comparison is significant at the 0.05 level. The randomization test P-values and the adjusted P-values for the six comparisons of treatments A, B, C, and D follow:

Comparison	A—B	A—C	A—D	B—C	B—D	C—D
P-value	0.021	0.007	0.017	0.047	0.006	0.134
Adjusted P-value	0.126	0.042	0.102	0.282	0.036	0.804

Only the A—C and the B—D treatment comparisons provide an adjusted significance value as small as 0.05, and so those are the only cases where H_0 of identical effects for a pair of treatments would be rejected at the 0.05 level.

To use Fisher's modified LSD procedure, we set an experiment-wise error rate or level of significance, such as 0.05 or 0.01, compute the significance value of t by data permutation for each of the $(k)(k-1)/2$ pairs of treatments, and then adjust those values by multiplying each value by $(k)(k-1)/2$ to get the final adjusted significance values. Unlike some multiple comparisons procedures, this procedure can be used either with equal or unequal sample sizes.

Any multiple comparisons procedure should be given careful consideration to determine whether it is needed, because the significance values are drastically affected by such procedures. Notice that, in Example 4.6, if the adjustment in the P-values had not been made there would have been five comparisons, instead of just two, with significance at the 0.05 level. It should also be noted that no unadjusted significance value could lead to an adjusted value of 0.05 or less, unless it was smaller than 0.01. When there are more than four treatments the adjustment has even greater effect on the P-values. With 10 treatments, for example, there are 45 possible treatment pairs, and so an unadjusted significance value would have to be about as small as 0.001 to provide significance at the 0.05 level for the final, adjusted value. Some multiple comparisons procedures alter the significance value more than others, but the effect of Scheffe's procedure, Tukey's "honestly significant difference" procedure, and other relatively uncontroversial procedures is comparable to that of Fisher's modified LSD procedure. Consequently, multiple comparisons procedures should be used only when necessary. If logic or editorial policy requires the use of multiple comparisons procedures they must be used, but there are cases where use of a few planned comparisons makes more sense. If multiple comparisons procedures are to be employed, the researcher should realize that with a large number of treatments he will need more subjects per treatment than with a small number of treatments to achieve the same sensitivity or power for individual comparisons.

There is a possibility of confusing Fisher's modified LSD procedure with an earlier procedure which is not appropriate for multiple comparisons; so in referring to the procedure, it should be specified as the *modified least significant difference* procedure.

4.12   LOSS OF EXPERIMENTAL SUBJECTS

It has been noted frequently in experimental design books that a statistical test can become invalid if the treatments are such that there is differential dropping out, in the sense that subjects might drop out if assigned to one treatment but not if assigned to another. The term "dropping out" refers to the case where a subject assigned to a treatment does not participate

enough to provide a measurement for that treatment. For
instance, a rat may be unable to finish an experiment because
it becomes sick or dies or does not reach a certain criterion
during the training trials. Whenever some subjects originally
assigned to treatments do not provide measurements under those
treatments, it is necessary to consider whether such dropping
out biased the statistical test carried out on the remaining subjects. That would be the case, for example, if a surgical
treatment caused the weaker rats to die, leaving only the
stronger rats in the group to be compared in respect to physical activity with rats taking another treatment.

There are instances, however, where we are quite sure that
the subjects that are lost from the experiment would also have
been lost if they had been assigned to a different treatment.
In such cases it is clear that the disease or other reason for
dropping out of the experiment would have existed even if the
subject had been assigned to one of the alternative treatments.
We then feel confident that there is no bias resulting from some
assigned subjects not finishing the experiment. We must go
further, though, and consider how the data permutation is to
be performed when some assigned subjects do not provide data.
It will be assumed in the following example that the subjects
that were lost after assignment to treatments would have been
lost if they had been assigned to any of the other treatments.

Example 4.7

We have three treatments and have randomly assigned five subjects to each treatment. One subject assigned to A and one
assigned to B dropped out, and so we are left with data from
only 13 of the subjects, as shown in Table 5. Although we
have only 13 subjects, we must consider the possible assignments for all 15 subjects to perform data permutation. There
are 15!/5!5!5! possible assignments. For each of those possible
assignments there is a data permutation based on the null hypothesis and the assumption that the dropouts (indicated in
Table 5 by dashes) would have dropped out wherever they were
assigned. The data permutation in Table 6, for instance, is
associated with one of the assignments where the two subjects
that dropped out are represented as having been assigned to C.
The data permutations would not, then, have the same sample
size allocations of *data* from one data permutation to the next.
Although it would be valid to compute F on the basis of such
data permutations, the variation in sample size allocations of

TABLE 5   Experimental Results

A	B	C
5	8	4
8	8	7
6	12	9
9	10	8
—	—	11

data from one data permutation to the next would prevent
determination of significance by use of the computer programs
given in this chapter.

An alternative procedure that permits use of the standard
programs is better. The data can be fed into the computer
as though the random assignment had been of four subjects to
A, four to B, and five to C, and the data would be permuted
accordingly, with no variation in sample size allocation from
one data permutation to the next. Instead of 15! /5! 5! 5! data
permutations being involved, there would be 13! /4! 4! 5!, a
subset of the total number, consisting of those data permutations corresponding to possible assignments when one of the
two subjects that dropped out was assigned to A and one to B.

TABLE 6   Data Permutation

A	B	C
5	8	4
8	8	7
6	12	9
9	10	—
8	11	—

The P-value is the proportion of the data permutations in the subset, of which the obtained data permutation is a member, that provide as large a value of F as the obtained data. When H_0 is true, the probability that the obtained F will be within the upper p percent of the F's in the subset is no greater than p percent, so this procedure of determining significance is valid. The computer programs in this chapter, therefore, can be validly employed for determining significance when there has been a loss of subjects after assignment (if the subjects would have been lost under any alternative assignment) by treating the data in the same way as if the data came from an assignment where all subjects provided data.

4.13   RANKED DATA

When more precise measurements are available, it is unwise to degrade the precision by transforming the measurements into ranks for conducting a statistical test.   This transformation has sometimes been made to permit the use of a nonparametric test because of the doubtful validity of the parametric test, but it is unnecessary for that purpose since a randomization test provides a significance value whose validity is independent of parametric assumptions, while using the raw data.   For a person determining significance by a randomization test there is no object in transforming the data into ranks for the statistical analysis.   However, there are occasions when the only available data are rankings, and for those occasions the t test and ANOVA programs used for ordinary data can be used on the ranks to determine significance.

An alternative to using data permutation to determine significance for ranked data for the independent t test is to use the Mann-Whitney U test, which has significance tables.   The significance tables are based on the permutation of ranks.   The test statistic U and alternative test statistics, like the sum of ranks for one of the treatments, are equivalent test statistics to t for ranks, given data permutation.   Thus, for ranked data *with no tied ranks*, reference to the U tables will give the same significance value as would be obtained by determining the significance of t (for the ranks) by means of a randomization test using systematic data permutation.   The Mann-Whitney U tables have been constructed for a wide range of sample sizes, and the only conceivable advantage in using data permutation on ranks rather than to use the U tables is in analyzing ranks

where there are a large number of tied ranks; since the significance tables for U are based on permutation of ranks, with no
tied ranks, they are only approximately valid when there are
tied ranks.

The Kruskal-Wallis analysis of variance procedure is a nonparametric procedure for testing the significance of difference
between two or more independent groups, and the published
significance tables for the test statistic H are based on permutation of ranks for small sample sizes. H is an equivalent test
statistic to F for one-way ANOVA under data permutation of
ranks; so for ranked data *with no tied ranks*, reference to the
H tables will give the same significance value as would be obtained by using a randomization test to determine the significance of F by systematic data permutation. However, for small
sample sizes where the significance values shown in the table
have been determined by permutation of the ranks, some tables
(e.g., Siegel, 1956, pp. 282–283) show significance values for
no more than three groups. For larger samples, the significance tables for H are not based on data permutation but on the
chi-square distribution and thus provide only an approximation
to the P-value that would be given by data permutation. Consequently, to find the significance of F for ranked data, data
permutation can usefully complement the H tables, for both small
and large samples.

4.14   DICHOTOMOUS DATA

It is not necessary to have a quantitative dependent variable
in order to use the computer programs in this chapter. Provided the random assignment has been conducted in the standard way for independent-groups experiments, the programs can
be used to determine significance by data permutation even for
a qualitative, dichotomous dependent variable. For instance,
a response may be simply designated as falling into one of two
categories (such as "lived" or "died" or "correct" or "incorrect")
but still permit the use of the t test or ANOVA. This is made
possible by assigning a value of 1 to every response in one of
the categories and a value of 0 to every response in the other
category. The 0's and 1's are the data for the experiment
which are permuted to determine the significance of t or F, as
the case may be. Suppose we assigned a 1 to each correct
response and a 0 to each incorrect response. Then the difference between treatment means to which the t test and ANOVA

are sensitive is a difference between the proportion of correct
responses for the different treatments, because the mean of the
0's and 1's for a treatment is simply the proportion of correct
responses for that treatment. Thus, by assigning 0's and 1's
to the two types of responses, we are able to carry out a meaningful test on the relative frequency of correct responses over
treatments.

Example 4.8

We have eight rats that are randomly assigned to three treatments, with three rats assigned to A, three to B, and two to
C. After receiving the treatments, all A rats made incorrect
responses, both C rats made correct responses, and two B rats
made a correct response and one made an incorrect response.
For ANOVA, the data would appear in the following way:

A	B	C
0	1	1
0	1	1
0	0	

There would be $8!/3!3!2! = 560$ permutations, because there
are that many possible assignments. Under H_0, a rat that made
a correct response would have made that response under any of
the treatments, and a rat that made an incorrect response would
have been incorrect under any of the treatments; thus, under
H_0, wherever a rat was assigned, its "measurement" of 0 or 1
would go with it. The significance of F would be the proportion
of the 560 data permutations that provide such a large value as
the F value for the obtained results.

Thus the programs for t and F in this chapter can be employed with dichotomous data, by assigning 0's to responses in
one category and 1's to responses in the other category, and
by determining the significance value for F or t. It would be
simpler to compute contingency χ^2 and find the significance in
the chi-square table, but chi-square tables only approximate
the significance given by data permutation. When there are
small expected frequencies, and just one degree of freedom,
one should "correct for continuity" by using the alternative

chi-square formula: $\chi^2 = \Sigma[(|o - e| - 0.5)^2/e]$. The absolute difference between the observed and expected frequencies is reduced by 0.5 before being squared, reducing the chi-square value and making it more conservative than the value given by the standard chi-square formula. Even with the correction for continuity, the significance value of chi-square as given in chi-square tables corresponds only approximately to the value given by data permutation. Without correction for continuity, the discrepancy may be considerable even when the expected values are relatively large.

A second reason for using data permutation with 0's and 1's to determine significance is that with only two treatment groups, one-tailed tests can be carried out, whether the sample sizes are equal or unequal. Chi-square tables, on the other hand, provide only two-tailed significance values and, therefore, are not as sensitive when a directional prediction is relevant.

There is a procedure called *Fisher's exact test*, which gives the same one-tailed significance value as data permutation, although it determines the value by direct computation, but the computation can sometimes be considerable without computational aids. Furthermore, Fisher's exact test does not give two-tailed P-values and is restricted to two treatments.

4.15   OUTLIERS

Outliers are extremely high or extremely low measurements, and are so called because they lie far outside the range of the rest of the measurements. When the outlier is a low measurement, it tends to produce a negative skewness in the distribution of data; when it is a high measurement, a positive skewness is produced.

It might seem that the addition of a high outlier to the set of measurements with the higher mean, or a low outlier to the set with the lower mean, would tend to increase the value of t since it increases the difference between means, but that is not necessarily the case. The addition of such an outlier can increase the within-group variability so much that it more than offsets the increase in the between-group difference, actually causing the t value to decrease.

Consequently, the presence of an outlier, even when it increases the difference between means, can sometimes reduce the value of t and thereby reduce the significance based on t tables. Since significance determined by a randomization test

is dependent upon the size of the obtained t value relative to
its size under other data permutations, and not on the absolute
size, the presence of outliers does not have the same amount
of depressing effect on randomization test significance. Randomization tests, therefore, may be more likely than t tables to
detect differences when there is a very extreme measurement.

Example 4.9

We will now illustrate this point with actual data from an experiment conducted by Ezinga (1976) which has been discussed
similarly elsewhere (Edgington and Ezinga, 1978). There were
10 subjects for each of two treatments and an independent t
test was used as a two-tailed test. Table 7 shows the data
and the significance given by the t table, along with the significance given by random data permutation, using 9,000 data
permutations. Although the P-value given by a randomization
test was 0.026, the results according to the t table would not
be significant at the 0.05 level. (The exact P-value based on
the parametric t distribution is 0.092.)

The effect of the outlier, 3.72, is so great that even if all
of the small measurements were in treatment A and all of the
large ones in treatment B, with no overlap, the t value would
be only 2.61, which would still not be significant at the 0.01
level. (The exact P-value based on the parametric t distribution is 0.018.) On the other hand, the significance value given
by a randomization test would be about 0.00001.

The data in Table 7 is just one of 11 sets with extreme
outliers. Eleven sets of data out of 71 sets on which an independent t test was conducted were so skewed that the presence
of the highest measurement doubled the range of the measurements. They all showed randomization tests to be more powerful for determining significance than the t distribution underlying the t table. Several dependent variables were used in the
overall study, which concerned memory for pictures. The study
required subjects to indicate whether a picture had or had not
been presented previously. None of the sets of data involving
accuracy measures showed extreme outliers. Since the subjects
presumably were strongly motivated, one would not expect
extreme differences between the best subject and the next
lower, or between the worst subject and the next higher. The
11 distributions that were extremely skewed due to outliers all
involved dependent variables where no ability was required to
obtain a high measurement. For example, some of the outlier

TABLE 7 Significance of t
for a Skewed Distribution

Treatment A	Treatment B
0.33	0.28
0.27	0.80
0.44	3.72
0.28	1.16
0.45	1.00
0.55	0.63
0.44	1.14
0.76	0.33
0.59	0.26
0.01	0.63
$\overline{X}$ = 0.412	0.995

Independent t test: t = 1.78,
18 degrees of freedom.

Significance by t table: between 0.05 and 0.10 (two-tailed).

Significance by randomization test: 0.026 (two-tailed)

Hypothetical case of maximum difference (no overlap).

Independent t test: t = 2.61,
18 degrees of freedom.

Significance by t table: between 0.01 and 0.02 (two-tailed)

Significance by randomization test: 0.00001 (two-tailed).

Source: Data from Ezinga's (1976) study.

distributions were distributions of confidence ratings made by
subjects to indicate their confidence in the accuracy of their
judgements. Since an occasional subject was extremely confident of his accuracy (whether justifiably so or not), such
subjects had confidence scores much larger than the rest.
Some dependent variables, then, by their very nature are likely
to lead to outliers, and in such cases the use of data permutation for determining significance may provide much smaller
P-values than t tables.

It is not the intent of the preceding discussion to imply
that randomization tests always will provide smaller P-values
than t tables for determining the significance of t when there
are outliers. Undoubtedly, the relative sensitivity of the two
methods of determining significance is a function of such considerations as the number of outliers and whether they fall
mainly at one end of the distribution or are evenly divided between the two ends. Much needs to be investigated before the
conditions under which randomization tests are more powerful
for data with outliers can be specified. The discussion does
show, however, that randomization tests not only control for
Type 1 error but, when there are treatment effects, also can
show greater power than conventional means of determining
significance for the same test statistic. Furthermore, it illustrates the fact that conditions which adversely affect significance
determined on the basis of parametric distributions of test statistics need not have a comparable effect on randomization test
P-values because of the randomization test's use of the *relative*
rather than the *absolute* value of a test statistic.

Outliers sometimes are identified and dropped from the data
set before performing a statistical test. One means is to compute
the mean and standard deviation of the joint distribution of
measurements and discard all measurements more than, say,
three standard deviations from the mean. The rationale for
discarding outliers seems to be that those measurements are in
some sense flawed. Whether one should rely on such criteria
to decide what data to analyze depends on one's assumptions
about the way "correct" measurements should be distributed.
Such a procedure of discarding outliers, if otherwise regarded
as appropriate, can be employed with a randomization test by
initially discarding outliers and then permuting the remaining
data, since under H_0 the same observations would be discarded
no matter which data permutation represented the obtained
results. Discarding an outlier on the basis of its extremeness

within a data set for an individual treatment, however, rather than on the basis of the joint distribution of measurements from all treatments, would be unacceptable for a randomization test procedure.

4.16   REPORTING RESULTS OF STATISTICAL TESTS

It is important for the reader of a research report with data analyzed by a randomization test to realize that the standard ANOVA or t test was run and that only the way of determining significance is different.   This can be accomplished by mentioning the randomization test only where the analysis of the data is discussed.   Throughout most of the report, reference can be made to the test conducted and the significance obtained without mentioning a randomization test.

In the analysis section, however, it is informative and to the experimenter's advantage in forestalling questions of statistical validity to explain that significance was determined by a randomization test.   To serve as a reminder of the validity provided by the randomization test, the statement of use of a randomization test should be accompanied by the reason for using it.   Other explanations than the nonrandomness of sampling may be given.   For instance, it could be explained that, although any of 10 numerical values on a rating scale could have been used, in fact only three values at one end of the scale accounted for practically all of the data, making the distribution too discrete to justify the assumption of continuity that underlies the t table.   To ensure validity the significance of t, therefore, was determined by use of a randomization test.

The term "randomization test" is a standard expression that is well known and there is no need to explain it in a presentation of research results.   On the other hand, if random permutation is employed, the number of data permutations used should be specified.

A technical consideration of importance in writing computer programs, but which is irrelevant in interpreting the results of the statistical test, is the use of an equivalent test statistic instead of t or F.   Since the P-values are the same as would have been obtained if the programs explicitly used t and F as test statistics, the experimenter should, of course, refer to the significance of t or F instead of, for example, the significance of the sum of measurements for treatment A or the significance of $\Sigma(T^2/n)$.

For many journals it is conventional to report not only the significance value for t or F but also values of the test statistics and the associated degrees of freedom. The experimenter can incorporate into the randomization test programs the computation of the conventional test statistic for the experimental results, or it can be computed separately. Wherever one would normally mention the conventional test statistic value and the degrees of freedom in a report, it can be done in this manner: "The response times for the two treatments were significantly different (t = 3.87, 10 degrees of freedom, p = 0.007 by randomization test)."

REFERENCES

Edgington, E. S., and Ezinga, G. (1978). Randomization tests and outlier scores. *J. Psychol. 99*, 259–262.

Ezinga, G. (1976). *Detection and Memory Processes in Picture Recognition*. (Ph.D. thesis). University of Calgary, Alberta.

Fisher, R. A. (1936). The coefficient of racial likeness and the future of craniometry. *Journal of the Royal Anthropological Institute, 66*, 57–63.

Green, B. F. (1963). *Digital Computers in Research*. McGraw-Hill, New York.

Keppel, G. (1973). *Design and Analysis: A Researcher's Handbook*. Prentice-Hall, Englewood Cliffs, NJ.

Miller, R. G. (1966). *Simultaneous Statistical Inference*. McGraw-Hill, New York.

Myers, J. L. (1966). *Fundamentals of Experimental Design*. Allyn and Bacon, Boston, MA.

Pitman, E. J. G. (1937). Significance tests which may be applied to samples from any populations. *Journal of the Royal Statistical Society Supplement, 4*, 119–130.

Siegel, S. (1956). *Nonparametric Statistics for the Behavioral Sciences*. McGraw-Hill, New York.

Winer, B. J. (1971). *Statistical Principles in Experimental Design*. (2d ed.). McGraw-Hill, New York.

5
Repeated-measures Analysis of Variance and the Correlated t Test

The use of randomization tests for determining significance for repeated-measures experiments is the topic of this chapter. The two principal tests used for such experiments are repeated-measures ANOVA and the correlated t test. Like one-way ANOVA and the independent t test, these tests are basic components of factorial and multivariate ANOVA, and so the programs in this chapter are relevant also to later chapters.

All subjects take all treatments in repeated-measures experiments, and so there is no need to control for between-subject variability. Random assignment provides statistical control over within-subject variability and is carried out in the following manner. Each subject has k treatment times, and the times are randomly assigned to the k treatments. The random assignment is carried out separately for each subject. There are thus $(k!)^n$ possible assignments, since the k! possible assignments for each subject can be associated with that many different assignments for every other subject.

Although repeated-measures experiments traditionally are discussed in terms of a number of subjects, each of whom takes each treatment, the programs in this chapter can be used also in single-subject experiments to determine significance. Such applications will be described in Chapter 10.

5.1   THE POWER OF REPEATED-MEASURES TESTS

The repeated-measures test is sensitive to the same kind of treatment effects as the independent t test or simple one-way ANOVA:
effects wherein some treatments tend to provide larger measurements than others. Thus, an experimenter would not employ a
repeated-measures test to look for a different kind of effect from
that to which a comparison between independent groups (groups
employing different subjects for different treatments) is sensitive.
The repeated-measures test is simply a more sensitive test than
the other kind when there is at least a moderate degree of positive correlation between measurements under the various treatments. To see why repeated-measures tests, of which the correlated t is the special type employed for the two-treatment
experiment, are likely to be more powerful when there is a positive correlation, let us first examine the question graphically.
Figure 1 shows two sets of hypothetical data. Set A consists
of the same measurements as those shown in set B; but set A
measurements are intended to represent the measurements of 10
subjects, five of which were assigned to each of the two treatments, whereas set B measurements represent the measurements
of only 5 subjects, each of which took both treatments. The
lines in set B connect the pairs of measurements generated by
the same subject. Given this explanation consider which graph
is more suggestive of a treatment effect. Certainly the overlap
of distributions for set A does not suggest very strongly a
differential effect of the two treatments. But although the measurements in set B have the same overlap, the lines that connect
the measurements of the same subjects indicate that there is a
consistent directional difference in measurements of about the
same magnitude in all subjects, and this fact leads one to sus-
pect a difference in the effects of the two treatments.

Repeated-measures tests deal with the *differences* between the
effects of treatments within individual subjects, and where these
differences tend to be consistent from subject to subject even
small differences among treatments can provide significant results.
In other words, when the variability within treatment groups is
large, a large difference among means is required to provide
significant results for one-way ANOVA, whereas with the same
variability within treatments the repeated-measures test may
find a small difference among means to be significant if there is
small variability within the *difference scores.*

As with the t test for differences between independent groups,
the t test of the repeated-measures kind (which sometimes is

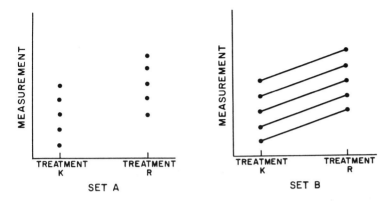

FIGURE 1   Graphs of two sets of data.

called the correlated t test) is simply a special case of a corresponding ANOVA test, and the squared t value equals the
value of F that repeated-measures ANOVA would give for the
same data.   Reference to the published probability tables would
provide the same probability values for t and F.   It is considered
desirable, however, to present both the t test and ANOVA for the
repeated-measures designs because a person may be interested
in the one-tailed P-value and such a P-value is available only
for the t test.

    There are fewer procedures to be considered here in connec-
tion with repeated-measures designs, because the standard repeated-measures designs use the same subjects for all treatments,
providing an equal-sample-size design.

5.2   SYSTEMATIC LISTING OF DATA
PERMUTATIONS

Since the random assignment procedure is different from that of
independent-groups experiments, repeated-measures experiments
require a different procedure of listing permutations to determine
significance by systematic data permutation.

Example 5.1

We have three subjects, each to be tested under two conditions,
A and B.   We could simply administer treatment A to all three

subjects and obtain measurements, then administer B to all subjects and obtain measurements. Then we could pair the measurements subject by subject and, no doubt, we would find that
within at least some of the pairs of measurements there was a
difference. That is, at least one of the subjects responded
differently to one of the treatments than to the other treatment.
But that does not answer the question of interest, which concerns a treatment effect. After all, from one treatment administration time to another, a subject may give different responses
to the *same* treatment, but we are not interested in the within-
subject variability that is simply a function of variation over
time. We are interested in the variability produced by the treatment differences. For this reason we should not give one treatment to all subjects first, then the other treatment. Instead,
we should randomly assign treatment times to treatments. For
each subject we have the treatment times, times that are convenient to both the experimenter and the subject. The earlier of
the times is called t_1 and the later t_2, and two slips of paper
are marked with those designations. One of the paper slips is
randomly selected for the first subject and that slip indicates
the treatment time for treatment A, the other treatment time being
associated with treatment B. The slip is replaced, then a drawing is made for the second subject to tell which treatment time
is associated with treatment A, and the same is done for the
third subject. The number of possible assignments is 2^3, or
eight, since there are two possible assignments for the first subject, each of which could be paired with two possible assignments
for the second subject, which in turn could be paired with two
possible assignments for the third subject. The experiment is
conducted and formula (3.1) is applied to the results to obtain t
for a correlated t test. The experimental results and the t value
are as follows:

	Treatments	
Subjects	A	B
a	(t_2) 18	(t_1) 15
b	(t_1) 14	(t_2) 13
c	(t_1) 16	(t_2) 14
	$\bar{X} = 16$	14
	t = 3.46	

In order to have a systematic procedure for listing data permutations, we first assign index numbers to the data and permute the index numbers systematically. Index numbers 1 and 2 are used to represent the data, where for each subject the measurement obtained under treatment A is given an index of 1, and the measurement obtained under treatment B is given an index of 2. Thus, the obtained results would be indexed in the following way:

Subjects	A	Index	B	Index
a	18	1	15	2
b	14	1	13	2
c	16	1	14	2

For purposes of listing a sequence of data permutations, the obtained results can be regarded as a six-digit number where the first two digits represent the index numbers for the first subject, the next two the index numbers for the second subject, and the last two the index numbers for the third subject: 121212. Each data permutation will be represented as a six-digit number where the digits 1 and 2 appear in each of the three groups of two digits in the six-digit number. Each successive data permutation provides a six-digit number that is next larger than the previous one. For example, the second permutation would be represented by 121221, the order of index numbers for the third subject being changed while the order for the first two subjects is held constant. For a complete listing of the data permutations, the six-digit number is incremented until it reaches its maximum value of 212121 on the eighth data permutation.

The eight data permutations are listed in sequential order in Table 1. The significance associated with the obtained results (which is represented by permutation 1) for a one-tailed t test where $\bar{X}_A$ is expected to be the larger mean, is 1/8, or 0.125, because only one of the eight data permutations provides a t value as large as the obtained value, 3.46. If no direction of difference were predicted, the significance would be determined for a two-tailed t test, and in that case the significance of the results would be 2/8, or 0.25, because two of the eight data permutations (permutations 1 and 8) provide |t|'s as large as the obtained value.

If only the subset consisting of the first four listed permutations in Table 1 was used to determine significance, the
same two-tailed P-value would be obtained. For a two-tailed
test, the last four data permutations are redundant since each
is a mirror image of one of the first four. Permutations 1 and
8 are alike except for transposition of the treatment designations,
which does not affect the t value, and the same is true of permutations 2 and 7, 3 and 6, and 4 and 5.

5.3   A NONREDUNDANT LISTING PROCEDURE

When the permutations are listed according to the rule in Section 5.2, it always will be the case for two treatments that the
second half of the data permutations will be redundant for
determining two-tailed significance. The first half of the data
permutations are those for which the measurements for the first
subject are in the obtained order; with all data permutations in
the first half having the same order for measurements for the
first subject, none can be a mirror image of another. Obviously,
in the set of all data permutations each data permutation can be
paired with another that is its mirror image. Consequently, for
each data permutation in the first half of the set, there is a
corresponding mirror-image permutation in the second half that
will give the same $|t|$ value, and so the two-tailed P-value
given by the subset consisting of the first half of the data permutations is necessarily the same as that based on the entire
set.
     With three treatments, because there are six ways the treat-
ment designations can be assigned to a particular division of
data, the data permutations can be grouped into sixes, where
each of the six is a mirror image of the others in that group.
Using the same sort of listing procedure for permutations, only
the first one-sixth of the data permutations need to be listed
to determine the significance of F, because they are the ones
that have the measurements for the first subject in the obtained
order. In general, then, for k treatments, only the first $1/k!$
of the data permutations need be computed to determine the
significance of F or $|t|$. Program 5.1 (see page 110) stops
performing data permutations when the next permutation would
permute the scores of the first subject, thereby ensuring that
the appropriate subset of $1/k!$ of the data permutations has
been used for determining significance.

TABLE 1 Eight Data Permutations

Permutation 1		Permutation 2		Permutation 3		Permutation 4	
A	B	A	B	A	B	A	B
18	15	18	15	18	15	18	15
14	13	14	13	13	14	13	14
16	14	14	16	16	14	14	16
$\bar{X} = 16$	14	$\bar{X} = 15\ 1/3$	14 2/3	$\bar{X} = 15\ 2/3$	14 1/3	$\bar{X} = 15$	15
t = 3.46		t = 0.46		t = 1.11		t = 0	

Permutation 5		Permutation 6		Permutation 7		Permutation 8	
A	B	A	B	A	B	A	B
15	18	15	18	15	18	15	18
14	13	14	13	13	14	13	14
16	14	14	16	16	14	14	16
$\bar{X} = 15$	15	$\bar{X} = 14\ 1/3$	15 2/3	$\bar{X} = 14\ 2/3$	15 1/3	$\bar{X} = 14$	16
t = 0		t = -1.11		t = -0.46		t = -3.46	

5.4 ΣT^2 AS AN EQUIVALENT TEST STATISTIC TO F

Table 2 is a summary table for repeated-measures ANOVA. In the table, k is the number of treatments and n is the number of subjects. $SS_{B(t)}$ is SS between treatments, $SS_{B(s)}$ is SS between subjects, and SS_e is the residual SS. $SS_{B(t)}$ and SS_e are defined as:

$$SS_{B(t)} = \sum \frac{T^2}{n} - \frac{(\Sigma X)^2}{kn} \tag{5.1}$$

$$SS_e = \Sigma X^2 - \sum \frac{T^2}{n} - \frac{\Sigma S^2}{k} + \frac{(\Sigma X)^2}{kn} \tag{5.2}$$

where k is the number of treatments, n is the number of subjects, T is the sum of measurements for a treatment, and S is the sum of measurements for a subject.

It will be shown now that ΣT^2 is an equivalent test statistic to F for repeated-measures ANOVA for determining significance by data permutation. The formula for F in Table 2 is equivalent to $(SS_{B(t)}/SS_e) \times [(n - 1)(k - 1)/(k - 1)]$. Because $(n - 1)(k - 1)/(k - 1) = n - 1$ is a constant multiplier over all data permutations, its elimination has no effect on the ordering of the data permutations with respect to the test statistic value. Thus $SS_{B(t)}/SS_e$ is an equivalent test statistic to F.

Reference to formulas (5.1) and (5.2) shows that all of the terms for the numerator and denominator of $SS_{B(t)}/SS_e$ except $\Sigma(T^2/n)$ are constant over all data permutations. An increase in $\Sigma(T^2/n)$ increases the numerator, shown in formula (5.1), and decreases the denominator, shown in formula (5.2), thereby increasing the value of $SS_{B(t)}/SS_e$. Therefore, $\Sigma(T^2/n)$ is an equivalent test statistic to $SS_{B(t)}/SS_e$, and thus is an equivalent test statistic to F. Each subject takes all treatments, and so $\Sigma(T^2/n)$ is the same as $(\Sigma T^2)/n$. With n constant over all data permutations, this is an equivalent test statistic to ΣT^2, the sum of the squares of the treatment totals. Because ΣT^2 is a simpler test statistic to compute for each data permutation than repeated-measures F and will give the same P-value as F, it will be used in the computer programs to determine the significance of repeated-measures F.

TABLE 2 Summary Table for Repeated-measures ANOVA

Source of variation	Sum of squares	Degrees of freedom	Mean square	F
Treatments	$SS_{B(t)}$	$k - 1$	$SS_{B(t)}/(k - 1)$	$\dfrac{SS_{B(t)}/(k - 1)}{SS_e/(k - 1)(n - 1)}$
Subjects	$SS_{B(s)}$	$n - 1$	$SS_{B(s)}/(n - 1)$	
Error	SS_e	$(k - 1)(n - 1)$	$SS_e/(k - 1)(n - 1)$	
Total	SS_T	$kn - 1$		

5.5   REPEATED-MEASURES ANOVA WITH SYSTEMATIC DATA PERMUTATION

The steps to take in determining significance of repeated-measures F by systematic data permutation are as follows:

Step 1. Arrange the data in a table with k columns and n rows, the columns indicating the treatments and the rows indicating the subjects. Assign index numbers 1 to n to the subjects and numbers 1 to k to the treatments, so that each measurement has associated with it a compound index number, the first part of which indicates the subject, and the second part of which indicates the treatment. For example, the measurement for the fifth subject under the treatment in column 4 would be given the index number (5,4).

Step 2. Compute GT, the grand total of the measurements over all treatments.

Step 3. Use a permutation procedure to produce a systematic sequence of permutations of the n measurements, where the measurements for the first subject are not permuted. That is, the measurements for the first subject are always associated with the treatments under which those measurements were obtained in the experiment, whereas measurements for the other subjects are permuted over treatments. Since exactly $1/k!$ of the $(k!)^n$ possible permutations are thereby performed, the number of permutations performed is $(k!)^{(n-1)}$, the first of which is the permutation represented by the observed data. Perform steps 4 to 6 for each of the permutations.

Step 4. Compute T_1, T_2, . . . , T_{k-1}, the total of the measurements associated with the index numbers for all but the last treatment.

Step 5. Compute $GT - (T_1 + T_2 + \cdots + T_{k-1})$ to get T_k, the total of the measurements associated with the index numbers for the last treatment.

Step 6. $T_1^2 + \cdots + T_k^2 = \Sigma T^2$ is the test statistic.

Step 7. Compute the probability for F as the proportion of the test statistic values that are as large as the obtained test statistic value, the test statistic value for the first permutation.

Example 5.2

The computational steps for this test will now be illustrated by a numerical example. Hypothetical data from an experiment are presented here according to the format prescribed in step 1:

Subject number	Treatment 1	Treatment 2	Treatment 3
1	10	11	13
2	13	14	17
3	6	8	9

The grand total of all the scores, specified in step 2, is 10 + 13 + 6 + 11 + 14 + 8 + 13 + 17 + 9 = 101. Going through steps 4 to 6 for the tabled data provides us with the obtained test statistic value. Since the totals for the first two treatments are 29 and 33, respectively, the total for the last treatment is 101 − (29 + 33) = 39. So the observed test statistic value is $(29)^2 + (33)^2 + (39)^2 = 3,451$. This value, 3,451, is the test statistic value for the first of the $(6)^2 = 36$ permutations that must be performed for the test. The second permutation is obtained by changing the sequence of scores for the third subject from 6 8 9 to 6 9 8. Then the test statistic ΣT^2 becomes $(29)^2 + (34)^2 + (38)^2 = 3,441$. There would be $(3!)^2 = 36$ permutations to perform, and the proportion of these permutations providing as large a test statistic value as the obtained value, 3,451, is the P-value associated with repeated-measures F for this example. The obtained data provides the largest test statistic value, as can be seen in the consistent increase from treatments 1 to 3 within each subject, and so the P-value is 1/36 = 0.028. If all 216 permutations had been used, the probability would be the same because there would be six orderings of the same columns of measurements that give the observed test statistic value. Since 6/216 is equal to 1/36, the probability obtained with 36 permutations is that which would be obtained with all 216 permutations.

Program 5.1 can be used to determine the significance of F by systematic data permutation. The comments in the program describe the modifications required to determine significance for

PROGRAM 5.1   Repeated-measures ANOVA:
Systematic Permutation

```
            ((INPUT))

      READ(5,---)N,NTRE
      TOTAL=NGE=NPERM=0
C*****FOR 1-TAILED T INSERT 'NGEONE=0'
      ITRE=NTRE-1
      DO 1 I=1,N
      DO 1 J=1,NTRE
      READ(5,---)DATA(I,J)
      TOTAL=TOTAL+DATA(I,J)
      INDEX(I,J)=J
      WITHOUT(I,J)=0
    1 IF(J.EQ.NTRE)WITHOUT(I,J)=1
    2 TEST=SUBTOT=0
C*****FOR 1-TAILED T INSERT 'TESTONE=0'
      DO 4 J=1,ITRE
      SUM=0
      DO 3 I=1,N
    3 SUM=SUM+DATA(I,INDEX(I,J))
      SUBTOT=SUBTOT+SUM
    4 TEST=TEST+SUM**2
      TEST=TEST+(TOTAL-SUBTOT)**2
C*****FOR 1-TAILED T INSERT 'TESTONE=TESTONE+SUM'
      NPERM=NPERM+1
      IF(NPERM.EQ.1)OBTAIN=TEST
C*****FOR 1-TAILED T INSERT 'IF(NPERM.EQ.1)OBTONE=TESTONE'
      IF(TEST.GE.OBTAIN)NGE=NGE+1
C*****FOR 1-TAILED T INSERT 'IF(TESTONE.GE.OBTONE)NGEONE=NGEONE+1'
      I=N
    5 J=ITRE
    6 WITHOUT(I,INDEX(I,J))=1
    7 IF(INDEX(I,J).EQ.NTRE) GO TO 8
      INDEX(I,J)=INDEX(I,J)+1
      IF(WITHOUT(I,INDEX(I,J)).EQ.0) GO TO 7
      WITHOUT(I,INDEX(I,J))=0
      IF(J.EQ.ITRE) GO TO 2
      J=J+1
      INDEX(I,J)=0
      GO TO 7
    8 J=J-1
      IF(J.GT.0) GO TO 6
      DO 9 J=1,NTRE
      INDEX(I,J)=J
    9 WITHOUT(I,J)=0
      WITHOUT(I,NTRE)=1
      I=I-1
C*****FOR 1-TAILED T CHANGE NEXT STATEMENT TO 'IF(I.NE.0) GO TO 5'
      IF(I.GT.1) GO TO 5
      PROB=FLOAT(NGE)/NPERM
C*****FOR 1-TAILED T INSERT 'PROBONE=FLOAT(NGEONE)/NPERM'

            ((OUTPUT))
```

---

a one-tailed correlated t test and are explained in Section 5.7.
Program 5.1 can be tested with the data given in Table 3,
which should give a P-value of $12/7,776 = 0.0015$.

Since the number of permutations depends not only on the
number of subjects but also on the number of treatments, the
number of permutations increases rapidly with an increase in the
number of subjects when there are several treatments. The
number of permutations to be computed is $(k!)^{(n-1)}$, and so
the addition of a subject multiplies the required number of permutations by $k!$. For three groups the addition of a subject
increases the number of permutations by a factor of 6, for four
groups by a factor of 24, and so on. For example, an experiment in which each of *three* subjects takes four different treatments involves a total of 13,824 permutations, one-twenty-fourth
of which need to be performed for the systematic permutation
test, and so it is feasible; but if *six* subjects were to take each
of four treatments, the total number of permutations would be
191,102,976, and one-twenty-fourth of the number still is too
many permutations for the determination of significance by systematic permutation to be feasible. Consequently, although determining significance by systematic permutation is useful for small
samples, it is not likely to be feasible for relatively large samples.
For relatively large samples, one should use the random permutation procedure which will be discussed in the following section.

5.6   REPEATED-MEASURES ANOVA WITH RANDOM
DATA PERMUTATION

The steps necessary in performing repeated-measures ANOVA
with random permutation are as follows:

Step 1. As with the systematic permutation procedure, arrange
the data in a table with k columns and n rows, where k is
the number of treatments and n is the number of subjects.
Assign index numbers 1 to n to the subjects and 1 to k to
the treatments, so that each measurement has associated
with it a compound index number, the first part of which
indicates the subject, and the second part of which indicates
the treatment. For example, the index (3,4) refers to the
measurement for the third person under the fourth treatment.

Step 2. Compute GT, the grand total of the measurements over
all treatments.

Step 3. Go through steps 5 to 7 for the obtained data to compute the obtained test statistic value.

Step 4. Use a random number generation algorithm that will randomly determine for each subject independently of the other subjects, which of the k measurements is to be "assigned" to the first treatment, which of the remaining k − 1 measurements to the second treatment, and so on. The random determination of order of measurements within each subject performed over all subjects constitutes a single permutation. Perform NPERM − 1 permutations, and for each permutation go through steps 5 to 7.

Step 5. Compute T_1, T_2, . . . , T_{k-1}, the total of the measurements associated with the index numbers for all but the last treatment.

Step 6. Compute GT − $(T_1 + T_2 + \cdot \cdot \cdot + T_{k-1})$ to get T_k, the total of the measurements associated with the index numbers for the last treatment.

Step 7. $T_1{}^2 + \cdot \cdot \cdot \cdot + T_k{}^2 = \Sigma T^2$ is the test statistic value.

Step 8. Compute the P-value for F as the proportion of the test statistic values, including the obtained test statistic value, that are as large as the obtained test statistic value.

Program 5.2 is the computer program for determining significance of repeated-measures F by random permutation. The comments in the program describe the modifications required to determine significance for a one-tailed t test and are explained in Section 5.8. Program 5.2 can be tested with the data in Table 3, for which the significance by systematic data permutation is 12/7,776 = 0.0015.

TABLE 3   Test Data

A	B	C
2	1	4
4	6	8
10	12	14
16	18	20
22	24	26

PROGRAM 5.2   Repeated-measures ANOVA:
Random Permutation

```
              ((INPUT))


       READ(5,---)NPERM
       READ(5,---)N,NTRE
C*****FOR 1-TAILED T INSERT 'NGEONE=0'
       TOTAL=NGE=0
       DO 1 I=1,N
       DO 1 J=1,NTRE
       READ(5,---)DATA(I,J)
     1 TOTAL=TOTAL+DATA(I,J)
       ITRE=NTRE-1
       DO 4 I=1,NPERM
C*****FOR 1-TAILED T INSERT 'TESTONE=0'
       TEST=SUBTOT=0
       DO 3 K=1,ITRE
       SUM=0
       DO 2 J=1,N
       IF(I.EQ.1) GO TO 2
       ID=K+RANF(0)*(NTRE-K+1)
       X=DATA(J,K)
       DATA(J,K)=DATA(J,ID)
       DATA(J,ID)=X
     2 SUM=SUM+DATA(J,K)
       SUBTOT=SUBTOT+SUM
     3 TEST=TEST+SUM**2
C*****FOR 1-TAILED T INSERT 'TESTONE=TESTONE+SUM'
       TEST=TEST+(TOTAL-SUBTOT)**2
C*****FOR 1-TAILED T INSERT 'IF(I.EQ.1)OBTONE=TESTONE'
       IF(I.EQ.1)OBTAIN=TEST
C*****FOR 1-TAILED T INSERT 'IF(TESTONE.GE.OBTONE)NGEONE=NGEONE+1'
     4 IF(TEST.GE.OBTAIN)NGE=NGE+1
C*****FOR 1-TAILED T INSERT 'PROBONE=FLOAT(NGEONE)/NPERM'
       PROB=FLOAT(NGE)/NPERM


              ((OUTPUT))
```

5.7   CORRELATED t TEST WITH SYSTEMATIC DATA PERMUTATION

For repeated-measures tests, as for independent-groups tests, F is equal to t^2, and so F and the two-tailed t test statistic $|t|$ are equivalent test statistics for determining significance by means of a randomization test. Thus, when only two treatments are employed, the P-value given by Program 5.1 is the value associated with a two-tailed correlated t test. Because of the symmetry of the complete sampling distribution of data permutations, halving the two-tailed P-value gives the one-tailed P-value where the direction of difference has been predicted *correctly*. For the purpose of combining probabilities, however, it is also useful to have a one-tailed probability value when the direction of difference is predicted *incorrectly*, and that value cannot be derived from the P-value given by Program 5.1 but must be determined separately by a program. For such one-tailed probability computation a special program for correlated t is necessary.

For correlated t as well as independent t tests, the difference between means can be computed in a manner such that a positive t is obtained when the difference between means is in the predicted direction and a negative t is obtained if the difference is in the opposite direction. Therefore we define the one-tailed probability for correlated t for a randomization test as the probability, under H_0, of getting such a large value of t (taking into consideration the sign of t) as the obtained value. (When the direction of difference between means has been correctly predicted, the one-tailed P-value computed in this manner will be the same as half of the two-tailed value given by Program 5.1.)

A one-tailed test statistic for correlated t that is equivalent under data permutation to t is T_L, the total of the measurements for the treatment predicted to give the larger mean. The following procedure for determining significance for one-tailed correlated t will use T_L as the one-tailed test statistic.

It is not necessary to write a completely new program for correlated t for determining one-tailed and two-tailed P-values by systematic data permutation, because a few changes in Program 5.1 will adapt it accordingly. After Program 5.1 has been modified, the first measurements put into the computer must be the measurements for the treatment predicted to provide the larger mean.

For Program 5.1 to give a one-tailed P-value for a correlated t test, regardless of whether the direction of difference
between means was predicted correctly, in addition to the two-
tailed P-values, the following changes should be made in the
programs:

1. *Change* IF(I.GT.1) GO TO 5 to IF(I.NE.0) GO TO 5
2. *Add* the following statements immediately *after* the corresponding statements for the two-tailed test:

```
NGEONE = 0
TESTONE = 0
TESTONE = TESTONE + SUM
IF(NPERM.EQ.1)OBTONE = TESTONE
IF(TESTONE.GE.OBTONE)NGEONE = NGEONE + 1
PROBONE = FLOAT(NGEONE)/NPERM
```

The *changed* statement is necessary to ensure that all, rather
than just half, of the data permutations are computed. The
required changes are specified in the comments in Program 5.1.

It must be stressed that the modified program uses the sum
of the measurements for the first listed treatment as the one-
tailed test statistic, and so it is essential that the first measurements entered into the computer be those for the treatment
predicted to provide the larger mean.

5.8   CORRELATED t TEST WITH RANDOM DATA
   PERMUTATION

With random permutation, as with systematic permutation, the
P-value for F for repeated-measures ANOVA when there are only
two treatments is the two-tailed P-value for correlated t. This
P-value can be divided by two to get the one-tailed P-value
for one-tailed t, whenever the direction of difference has been
predicted *correctly*. The justification of obtaining the one-tailed
P-value in this way is the same as in Section 4.9 for random
data permutation with independent t tests with equal sample
sizes: by virtue of the symmetry of distribution of measurements
within the complete sampling distribution, the probability is 1/2
that, in a random sample of data permutations, for any $|t|$ the
difference between means will be in the predicted direction when
H_0 is true. Thus the probability of getting a value of $|t|$ (or
the equivalent test statistic, ΣT^2) as large as an obtained value

with the obtained direction of difference between means, under
H_0, is half of the probability associated with two-tailed t.

Program 5.2, therefore, can be used to determine the two-
tailed P-value for correlated t, and half of that value is the
one-tailed value when the direction of difference between means
has been predicted correctly. When the direction of difference
has been predicted *incorrectly*, however, the one-tailed probability cannot be derived from the probability given by Program
5.2 but must be computed separately. Modifications in the
program similar to those suggested for Program 5.1 make it
appropriate for computing a one-tailed P-value for a correlated
t test, regardless of the correctness of the prediction of direction of difference between means. With these corrections the
program will provide both one-tailed and two-tailed P-values for
correlated t.

The modified program uses the sum of the measurements for
the first listed treatment as the one-tailed test statistic and,
therefore, assumes that the first measurements put into the computer are those for the treatment predicted to provide the larger
mean.

The only changes in the program that are required are the
following statements, which should be added to Program 5.2
immediately *before* the corresponding statements for the two-
tailed test:

```
NGEONE = 0
TESTONE = 0
TESTONE = TESTONE + SUM
IF(I.EQ.1)OBTONE = TESTONE
IF(TESTONE.GE.OBTONE)NGEONE = NGEONE + 1
PROBONE = FLOAT(NGEONE)/NPERM
```

The required changes are specified in the comments in Program
5.2.

5.9 CORRELATED t TEST AND PLANNED COMPARISONS

In Section 4.10 there was a discussion of the independent t test
and planned comparisons. The same sort of considerations apply
to planned comparisons with the correlated t test. With a repeated-measures experiment involving k treatments there are
$(k)(k - 1)/2$ pairs of treatments which could be subjected to a
correlated t test. When there are a small number of comparisons

planned before the experiment, the significance of the correlated t's for those comparisons can be determined by data permutation for each comparison as if the two treatments in the comparison were the only ones in the experiment. For instance, no matter how many treatments were administered in the experiment, for a planned comparison of treatments B and D the significance of t based on systematic data permutation would depend only on the 2^n permutations of the data for those two treatments. (H_0 would, of course, refer to treatments B and D alone.) Program 5.1 would provide the P-value for a two-tailed test for systematic data permutation or, with the modifications indicated by the comments in the program, it would provide the appropriate one-tailed P-value. If the number of data permutations is so large as to necessitate random permutation, Program 5.2 can be used for the two-tailed test. If Program 5.2 is modified according to the program comments, it will give the appropriate one-tailed P-value based on random permutation.

5.10 CORRELATED t TEST AND MULTIPLE COMPARISONS

In Section 4.11, the discussion concerning the use of the independent t test with multiple comparisons is in general applicable also to the correlated t test. Before the development of special procedures for multiple comparisons, it was customary first to determine the significance of the overall F for a repeated-measures analysis of variance, and then, if F was significant at the 0.05 level, to determine the significance for each of the possible comparisons by the correlated t test, with no adjustment of the significance values to take the number of comparisons into account. Such a treatment of posthoc comparisons, following determination of a significant F, still is used by some researchers; however, adjustment of the significance values to take into account the number of comparisons is more acceptable.

Fisher's modified least significant difference procedure, described in Section 4.11, is appropriate for multiple comparisons with repeated-measures as well as independent-groups data. The significance value of correlated t is computed by data permutation for each of the $(k)(k-1)/2$ pairs of treatments, where for each comparison only the measurements for the compared treatments are permuted. Next, those significance values are adjusted by multiplying each value by $(k)(k-1)/2$ to get the final adjusted significance values, as described in Section 4.11.

For repeated-measures experiments, as for experiments with independent groups, multiple comparisons procedures should be used only when necessary, because of their depressing effect on significance. A few thoughtfully planned comparisons may be a better approach. If a multiple comparisons procedure is appropriate, Fisher's procedure can be used. To avoid confusion, in reporting research results the procedure should be specified by its complete name: Fisher's modified least significant difference procedure.

5.11   RANK TESTS

Data permutation can be used to determine significance of correlated t or repeated-measures ANOVA for ranks as well as for raw data. Since it is as valid to use raw data with randomization tests as to use ranks, and since the use of raw data makes the test more sensitive to treatment effects, ranks should not be used when there are more precise measurements available.

Formula (3.1) for correlated t shows computation based on the difference scores. A data permutation procedure that would give identical P-values to the systematic procedure described in this chapter would be to perform permutations of the difference scores by assigning a + or − sign to the difference scores in all 2^n possible ways and using $|\overline{D}|$ as the two-tailed test statistic and $\overline{D}$ as the one-tailed test statistic.

Wilcoxon's matched-pairs signed-ranks test is the rank-order counterpart of the correlated t test. The absolute difference scores $|D|$'s are ranked over subjects and the ranks of $|D|$ are used in the analysis. Permuting the data consists in assigning a + or a − sign to the ranks of $|D|$ in all 2^n possible ways. The test statistic is the sum of ranks associated with the + signs. This is a widely used test for which significance tables are readily available.

If, instead of assigning a + or a − sign to the ranks of $|D|$, we assigned + or − signs to the $|D|$'s themselves and used the sum of the $|D|$'s associated with a + sign, the test would be equivalent to the randomization test for correlated t. But Wilcoxon's matched-pairs signed-ranks test is not like the randomization test applied to ranked data. The $|D|$'s in the Wilcoxon test that are ranked are absolute differences between measurements, not between ranks, and so the Wilcoxon test presupposes measurements that are more than ranks. Thus, the precision of the original measurements is degraded in order to free the test from parametric assumptions, a procedure which is no longer necessary when the randomization test procedure can be applied to the original measurements without any parametric assumptions. Transforming precise measurements to ranks for the purpose of employing a test free of parametric assumptions is no longer necessary, and as the Wilcoxon matched-pairs signed-ranks test necessarily involves such a transformation, it should be replaced by the randomization test for correlated t.

Suppose, however, that the original data was based on the ranking of responses to two treatments for each subject. If the responses for each subject are assigned ranks 1 and 2 and the 1's and 2's are entered as data in Program 5.1, the P-values obtained will be those associated with the sign test, which is a special application of the binomial test for which significance tables are published widely.

Situations arise in which one is able to rank two responses only within subjects, in which case the binomial test just described could be used. An example would be where the response was preference for one of two treatments (stimuli); while ranking of responses within subjects could be made, there would be no basis for ranking over subjects. But when ranking over subjects as well as within subjects is possible, the greater precision of such ranking should be used by employing an alternative to the binomial test. An example of ranking over subjects as well as within subjects would be ranking photographs consisting of two photographs from each of 10 subjects in order of attractiveness, where the pair of photographs of each subject is taken under two experimental conditions intended to influence facial expression. A reasonable test statistic would be the sum of ranks for one of the treatments. No significance tables for such a rank test exist because it would be unduly complex: instead of there being a single sum of ranks for significance at some specified level for n subjects, there would be as many sums of ranks as there are ways of dividing 2n ranks over n subjects with two ranks per subject. To use such a significance table, one would look up the appropriate n in the table and then look over all patterns of paired ranks to find the pattern associated with the experimental results and then find the sum of ranks for that pattern required for significance at the desired level. The lack of significance tables for this test is no problem, however, as the significance values that would be obtained from such tables can be directly obtained from Program 5.1 by entering all 2n ranks as the data.

There is no extension of the Wilcoxon matched-pairs signed-ranks test to more than two treatments, but the randomization test based on ranking over treatments and subjects described in the preceding paragraph is, of course, applicable to two or more treatments. There also is a multitreatment analogue of the sign test, which can be used when there are two or more treatments whose responses are ranked only within subjects. It is the nonparametric test called Friedman's analysis of variance by ranks.

The Friedman ANOVA-by-ranks test is a nonparametric test that uses ranked data to determine significance, and the significance values that are published for the Friedman test statistic are based on permutation of ranks. For each subject, the measurements over treatments are ranked and the rank numbers 1 to k are used as the measurements for the k treatments. The Friedman test statistic is equivalent to repeated-measures F under data permutation, and so the P-value obtained from the Friedman table is that which would be obtained for repeated-measures F under data permutation of the ranks. Program 5.1 with ranked data gives the same P-value as tables for the Friedman test statistic. Program 5.1 can be useful in extending the utility of Friedman's ANOVA, because of limitations of the tables. The Friedman table is, like other rank-order tables, derived from ranks with no ties, and so the significance values are only approximate when there are tied values. Also, some tables (see, e.g., Siegel, 1956, pp. 280—281) are restricted to nine subjects or less for three treatments and four subjects or less for four treatments. For larger samples chi-square tables must be used to provide estimates of the P-values based on data permutation.

Friedman's ANOVA uses ranking within subjects (or among matched subjects) only and thus wastes information when there is a basis for ranking over both subjects and treatments within subjects. The lack of significance tables for a rank test involving the use of kn ranks over both subjects and treatments need not handicap a person desiring such a test because the randomization test for repeated-measures ANOVA performed on the kn ranks is the desired test.

5.12   DICHOTOMOUS DATA

When the dependent variable is dichotomous, 0's and 1's can be assigned to the two categories, as with independent groups, and repeated-measures ANOVA can be used with the 0's and 1's to get a value of F whose significance is determined by data permutation.

The test statistic for the Cochran Q test (Siegel, 1956, pp. 161–166) is an equivalent test statistic to F for repeated-measures ANOVA, for determining significance by data permutation with dichotomous data. The Cochran test statistic is referred to chi-square tables to get an approximation to the P-value that would be provided by systematic data permutation: the value provided by Program 5.1.

5.13 COUNTERBALANCED DESIGNS

Let us consider next a procedure called "counterbalancing" that is sometimes used in repeated-measures experiments to control for within-subject variability. A method of *partial counterbalancing* is ensuring that each treatment occurs equally often in each serial position over all subjects. For example, given six subjects who each take three treatments, treatment A would be the first treatment taken by two subjects, B would be the first treatment for two others, and C would be the first treatment for the remaining two subjects. Similarly, each of the treatments would be taken second by two subjects and each of the treatments would be taken third by two subjects. Three sequences of treatments are sufficient for this purpose: ABC, BCA, and CAB. If two subjects take the ABC order, two take the BCA order, and two take the CAB order, each of the treatments will occur twice as the first treatment, each will occur twice as the second treatment, and each will occur twice as the third treatment. To provide adequate control over within-subject variability, there should be random assignment of the sequences to the subjects. This could be done in the following manner. Specify three treatment times for each subject; then designate the six subjects as a, b, c, d, e, and f on slips of paper that are placed in a container. The first two slips of paper that are randomly drawn from the container show the two subjects that take the treatments in the ABC order, the next two show the subjects that take the treatments in the BCA order, and the last two show the subjects that take the treatments in the CAB order. There are $6!/2!4! = 15$ possible pairs of subjects that can be selected for the ABC order, and for each of those possible pairs there are $4!/2!2! = 6$ possible pairs of subjects for the BCA order; but once the pair of subjects taking the ABC order and the pair taking the BCA order are determined, there is only one pair left to take the CAB order. Thus there are 15×6, or 90, possible assignments of subjects to treatment orders, and since the treatment order

for a subject assigns a treatment time to each treatment for that
subject, there are 90 possible assignments of treatment times to
treatments.

Example 5.3

Assume that subjects b and e were assigned to the ABC order,
subjects a and f were assigned to the BCA order, and subjects
c and d were assigned to the CAB order and that the experimental results shown in Table 4 were obtained. To determine
the significance of F, the equivalent test statistic ΣT^2 is computed for all 90 possible assignments of treatment times to treatments. The null hypothesis is that each subject gave the same
response at each of the treatment times as that subject would
have given if the treatments had been given in one of the other
two possible orders. So, under H_0, whatever the treatment
order assigned to a subject, the measurements associated with
the individual treatment times stay with their respective treatment times and this determines the data permutations associated
with each of the possible 90 assignments. The P-value is the
proportion of the 90 data permutations that provide as large a
value of ΣT^2 as 64,037, the obtained value.

The number of orders of treatments necessary for partial
counterbalancing is simply the number of treatments, since the
orders can be generated sequentially by taking the first listed
treatment and moving it to last place. For example, with five
treatments, five orders that allow each treatment to occupy each
serial position once are these: ABCDE, BCDEA, CDEAB,
DEABC, and EABCD. To have each treatment occur as often
as any other in each of the serial positions, it is necessary for
the number of subjects to be a multiple of the number of treatments. For example, with partial counterbalancing and five
treatments, 5 or 10 subjects could be used in the experiment
but not six or seven.

Complete counterbalancing is used by experimenters to provide better experimental control over within-subject variability
than can be provided by partial counterbalancing. Whereas
partial counterbalancing ensures that each treatment occurs
equally often in each serial position, complete counterbalancing
ensures that all possible serial orders of treatments occur equally
often. Not only would ABC, BCA, and CAB be used in an experiment with three treatments but also ACB, BAC, and CBA.
Each of the six possible orders would be used in the experiment
the same number of times. Thus, as with random assignment of

TABLE 4 Partial Counterbalancing

	Treatments		
Subjects	A	B	C
a	(t_3) 23	(t_1) 38	(t_2) 17
b	(t_1) 26	(t_2) 31	(t_3) 28
c	(t_2) 15	(t_3) 19	(t_1) 22
d	(t_2) 34	(t_3) 35	(t_1) 29
e	(t_1) 18	(t_2) 27	(t_3) 12
f	(t_3) 24	(t_1) 19	(t_2) 18
	$T^2 = 19,600$	28,561	15,876
	$\Sigma T^2 = 64,037$		

treatment times to treatments independently for each subject, all possible orders of treatments are considered; but unlike the independent assignment procedure the complete counterbalancing procedure makes sure that every possible order will be used and that each order will be used as often as any other.

To justify statistical analysis, complete counterbalancing should be done by a random assignment procedure. All k! orders of the k treatments would be listed, and then the first n randomly selected subjects, where n is the number of subjects per treatment order, would receive the first order, the second n would receive the second order, and so on. The minimum number of subjects for complete counterbalancing would be k! and, if more than k! subjects are used, the number must be a multiple of k! to ensure complete counterbalancing.

For a given number of subjects, then, partial counterbalancing, complete counterbalancing, and other procedures involving random assignment lead to different numbers of possible assignments, and so the P-value determination must follow the particular type of random assignment employed in the repeated-measures experiment. There will always be fewer permutations to consider with partial or complete counterbalancing than without counterbalancing. In unrestricted repeated-measures assignment, where the treatment times are randomly assigned for each subject independently of all other subjects, the distribution of permutations must include all those that would be obtained by complete counterbalancing, as well as additional permutations. The additional permutations are for cases where the number of subjects receiving the treatment in one order would not be the same as the number receiving the treatments in a different sequential order. If one counterbalances, then, when the sequence in which the treatments are given is likely to be of little consequence, the power of the test has been reduced because the total number of possible permutations is reduced. On the other hand, where it is felt that the sequence in which the treatments are given is of considerable importance, one should counterbalance. If the type of random assignment that is performed is taken into consideration in permuting the data, both counterbalancing and not counterbalancing are valid methods; but the relative power of the two methods depends on the importance of the sequence in which the treatments are given.

5.14   OUTLIERS

In Chapter 4 it was shown that data permutation may give substantially smaller P-values than t tables for the independent t test when there are outliers (extremely large or extremely small measurements). The presence of outliers for the correlated t test has a similar depressing effect on the power of the t table. The value of t for a difference between the means of treatments A and B depends on the magnitude of the difference between means relative to the variability of the difference scores. The addition of a subject with the same direction of difference as the means may increase the variability of the difference scores so much that it more than offsets the effect of the increase in the difference between means, causing the value of t to decrease.

Example 5.4

Ezinga's (1976) study was discussed in Chapter 4 in regard to the effect of outlier scores on the independent t test. He obtained outlier difference scores for correlated t tests also, which have been discussed elsewhere (Edgington and Ezinga, 1978). As might be expected, data permutation provided smaller P-values than the t tables for such data. Table 5 shows one data set which Ezinga analyzed by use of the correlated t test. The data are for 10 subjects evaluated under two alternative sets of instructions. The correlated t value is 1.97 which, for a

TABLE 5   Experimental Data Analyzed by Use of Correlated t Test

Subjects	New original instructions (A)	New repeated instructions (B)	Difference scores (B − A)
a	0.28	0.73	0.45
b	0.80	1.14	0.34
c	3.72	3.72	0.00
d	1.16	3.72	2.56
e	1.00	1.00	0.00
f	0.63	1.00	0.37
g	1.14	1.24	1.10
h	0.33	0.59	0.26
i	0.26	0.26	0.00
j	0.63	1.32	0.69

t = 1.97, 9 degrees of freedom

Two-tailed significance by t table: between 0.05 and 0.10

Two-tailed significance by randomization test: 0.016

two-tailed test, does not reach the 0.05 level in the t table; in fact, the parametric t distribution gives a P-value just larger than 0.08 for a t value of 1.97 with 9 degrees of freedom. Systematic data permutaiton, however, gives a two-tailed P-value of 0.016.

Data permutation gave more significant results than the t table for the data in Table 5, but is the difference a function of the outlier score? To answer this question, the data for subject d, who provided the outlier difference score of 2.56, were removed and the remaining data were analyzed by a correlated t test. With the outlier removed, the new t value was 3.05, which, with 8 degrees of freedom, has a P-value of 0.016 on the basis of the parametric t distribution. The direction of difference in scores for subject d is consistent with the direction for the other subjects, but the difference is so great that the inclusion of subject d depresses the significance of the results.

5.15 MATCHED-SUBJECTS DESIGNS

It is sometimes infeasible or even impossible to take repeated-measures on a subject. A swimming instructor, for instance, would not use a repeated-measures design to compare various methods of teaching the backstroke; once it has been learned it cannot be learned again. Some treatments have irreversible effects, making the random determination of order impossible. In a study of the performance of an animal before and after a brain lesion, random assignment of treatment times to "no lesion" and "lesion" conditions is impossible, and so there could be no statistical determination of treatment effect with a repeated-measures design.

Such situations called for a design where each subject takes only one treatment. Instead of using the assignment procedure specified in Chapter 4, however, we may choose to employ a more restricted assignment procedure, in which subjects are divided into groups and random assignment to treatments is carried out within each group. For instance, 10 subjects (each of which has a predetermined treatment time) are divided into five pairs of subjects in a way designed to make the two subjects within each pair very similar in characteristics relevant to performance. For the first pair we randomly select one of the two subjects to take treatment A at his predetermined treatment time, and the other subject is to take treatment B at the treatment time that was determined for him. The procedure is carried out for all five pairs of experimental units, where an experimental unit is subject-plus-treatment-time, providing a total of $2^5 = 32$ possible assignments, the number of assignments possible for five subjects with two treatments in an ordinary repeated-measures experiment. The data can be analyzed by means of Program 5.1 or 5.2 to determine significance by data permutation. The null hypothesis tested is: the measurement for each experimental unit is independent of the assignment to treatments. Since the experimental unit is subject-plus-treatment-time, the null hypoth-can be restated as: the measurement for each subject is the same as it would have been if the subject had taken the other treatment as his treatment time.

In general, given k treatments and n groups of k subjects each, there are $(k!)^n$ possible assignments, and significance can be determined by data permutation using Program 5.1 or 5.2. In using these programs, however, "number of subjects" for repeated-measures designs becomes "number of groups" and "data for subject i" becomes "data for group i."

Fisher's (1935) randomization test applied to Darwin's data on plant length, which was described in Section 1.13, was of the matched-subjects kind, as it involved random assignment of pairs of seeds to pairs of pots. Application of Program 5.1 to the data Fisher analyzed will give the same P-value as Fisher obtained. But Fisher's test required the assumption of random sampling from infinite populations to test the identity of the populations, whereas the principal value of Program 5.1 is its applicability to experimental data from experiments not involving random sampling. Pitman (1938) approached the randomization test for matched-subjects designs in a different manner from Fisher. He described the application of randomization tests to data from randomized block experiments generally, and the matched-subjects experiment in particular, where "the observed numbers are not really regarded as a sample from a larger population" (p. 322). In other words, Pitman described the same procedure for determining significance as Fisher, but Pitman assumed only random assignment.

The restriction on the random assignment imposed by a matched-subjects design provides such a reduction in the number of data permutations from the number associated with the same number of subjects for independent-groups assignment that one should carefully consider the use of independent-groups assignment before choosing a matched-subjects design. For example, with three subjects taking treatment A and three taking treatment B, there are 20 data permutations for independent-groups assignment, and so there is the possibility of a significance value as small as 1/20, or 0.05. However, use of three matched pairs of two subjects each yields only eight data permutations, and so the smallest possible significance value would be 1/8, or 0.125. In general it would be better to strive for homogeneity when selecting subjects for an experiment and to use an independent-groups design than to divide a heterogeneous sample of subjects into homogeneous groups for a matched-subjects design. The same rule applies here as applies to counterbalancing: make sure there is a good reason for restricting the random assignment procedure before placing severe constraints on it.

5.16 REPORTING RESULTS OF STATISTICAL TESTS

The reporting of results for repeated-measures ANOVA and correlated t tests follows the general guidelines given at the end of Chapter 4 for reporting results of conventional tests when significance is determined by means of randomization tests. The F or t value ordinarily would be reported, along with the associated degrees of freedom and the significance value, and it would would be indicated that the significance was determined by employing a randomization test. When counterbalanced designs are used, however, the report also should refer to the special type of random assignment that is employed and point out that the data were permuted for the randomization test in accordance with the special type of random assignment used.

REFERENCES

Edgington, E. S., and Ezinga, G. (1978). Randomization tests and outlier scores. *J. Psychol. 99*, 259—262.

Ezinga, G. (1976). *Detection and Memory Processes in Picture Recognition* (Ph.D. thesis). The University of Calgary, Alberta.

Fisher, R. A. (1935). *The design of experiments*. Oliver and Boyd, Edinburgh.

Pitman, E. J. G., (1938). Significance tests which may be applied to samples from any populations: III. The analysis of variance tests. *Biometrika, 29*, 322—335.

Siegel, S. (1956). *Nonparametric Statistics for the Behavioral Sciences*. McGraw-Hill, New York.

6

Factorial Designs

Factorial designs are designs for analyzing data where the data are classified according to more than one independent variable. The independent variables are called *factors*, the source of the name *factorial analysis of variance*, and the synonym *multifactor analysis of variance*, for a statistical procedure commonly used with such designs. A magnitude or state of an independent variable or factor is a *level* of the factor. For example, if we have an experiment in which plants are assigned to one of three amounts of humidity and to either high or low temperature, it is a factorial experiment with two factors (humidity and temperature), and there are three levels of the first factor and two levels of the second. In this case, both factors have levels which can be expressed quantitatively, but in some factorial experiments the levels of a factor are not magnitudes of the factor. For example, a factor studied in regard to plant growth could be the type of fertilizer, where four different kinds of fertilizer were compared; in this case there would be four levels of the fertilizer factor, although the variation over levels represents qualitative rather than quantitative differences.

Factorial designs are possible where none of the factors is manipulated, but factorial *experiments* require that at least one of the factors be manipulated; that is, that experimental units be assigned at random to the levels of at least one of the factors. The other factors also may be manipulated or they may be subject or organismic variables like sex or age.

One purpose of factorial designs is to increase the sensitivity of an experiment to treatment effects. Increased sensitivity to

treatment effects can result from stratifying subjects with regard
to a subject variable, as in a treatments-by-subjects design.
For instance, we may have one experimentally manipulated factor
and we may stratify our subjects into age levels that are relatively homogeneous, because we want to control for the heterogeneity in our data resulting from age differences.

A second purpose of factorial designs is to use the same
data to test the effect of more than one factor. If there are a
levels of one factor and b levels of a second factor in a two-
factor experiment, the number of *cells* in the matrix of treatment
combinations is ab, and each cell represents a distinctive treatment. Many tests are possible. The data can be used to test
separately the effects of factor A and factor B, or we can test
the effect of either factor within any single level of the other
factor. Thus, by using the same data for various comparisons,
we can make more complete use of our data than in a single-
factor design.

A third purpose of factorial designs is to permit experimenters to investigate interactions. An interaction is a difference in the effect of a factor from one level to the next level
of another factor. Especially interesting are *disordinal interactions*, interactions in which the rank order of levels of one
factor with respect to measurement means varies over the levels
of another factor.

6.1   ADVANTAGES OF RANDOMIZATION TESTS
FOR FACTORIAL DESIGNS

The validity of determining significance by use of randomization
tests when the parametric assumptions (including that of random
sampling) underlying F tables are not tenable is of more practical
importance with factorial analysis of variance (ANOVA) than with
the simpler procedures in Chapters 4 and 5. When the simpler
procedures are employed with continuous (not categorical) data
with relatively large and equal samples, the use of F tables to
determine significance seldom is called into question, even in the
absence of random sampling. On the other hand, under similar
conditions the use of factorial ANOVA may be challenged frequently because of the assumptions underlying the F table when
used with factorial designs. The assumptions are more complex,
and studies of the "robustness" of tests under violations of parametric assumptions have generally concerned the simpler ANOVA
procedures. Additionally, there frequently is need for a factorial

design when the sample sizes (in terms of the number of subjects per cell or per level of a factor) must be small or unequal,
and in those cases the validity is more likely to be questioned
than otherwise.

A randomization test procedure with considerable potential
for experimentation is a method of testing treatment effects
when the proportional distribution of cell frequencies over levels
of one factor varies from one level of a second factor to another.
One application of this randomization technique allows an experimenter to make effective use of subjects who are not willing to
be assigned to certain of the treatments. For each subject the
random assignment can be restricted to those treatments to which
the subject is willing to be assigned, and the restriction on the
possible assignments for a subject is taken into account in the
randomization test to permit a valid test of treatment effects.

6.2   FACTORIAL DESIGNS FOR COMPLETELY
RANDOMIZED EXPERIMENTS

First we will consider completely randomized factorial experiments,
that is, factorial experiments with random assignment to levels
of all factors. Random assignment for completely randomized
factorial experiments is of the same form as that for single-
factor experiments for one-way ANOVA. In fact, the data layout for a two-factor experiment in r rows and c columns could
be subjected to a randomization test for one-way ANOVA by regarding each of the rc cells in the array as a distinctive treatment rather than as a combination of treatments. To do so,
however, for testing the H_0 of no treatment effect whatsoever,
would be to lose certain advantages of a multifactor experiment
by reducing it to a single-factor (one-way) experiment. Advantages of factorial designs arise from conceptually dividing distinctive treatments (cells) into groups to permit useful comparisons of groups of treatments. When the desired organization
of data into factors and levels of factors is known in advance,
the experiment can be conducted more efficiently, but valid tests
of treatment effects can be conducted even when the grouping
of treatments for comparison is arranged after the experiment is
performed, provided the grouping is done before the data are
examined. For example, for practical purposes, an experiment
may be run to compare the effectiveness of reading a gauge as
a function of its location in one of four quadrants in front of a
machine operator. A completely randomized design, with ten

TABLE 1   Data Array

		HORIZONTAL	
		LEFT	RIGHT
VERTICAL	TOP		
	BOTTOM		

subjects assigned to each of the four conditions, is used. One-way ANOVA over all four cells could be employed to provide a single-factor analysis. A colleague with somewhat different interests, however, could analyze the experiment as a factorial experiment with the data for the four quadrants placed into the data array in Table 1 for two factors with two levels of each factor. The colleague carries out a test to compare column means to test the effect of horizontal placement of the gauge and then compares row means to test the effect of vertical placement. All four cells are involved in each of the tests, but what is tested is quite different from what is tested with a one-way ANOVA comparison of all four cells.

Null hypotheses for randomization tests for experiments that do not involve random sampling do not refer to population parameters; consequently, main effects and interactions will be defined here somewhat differently for randomization tests than for parametric tests. We will define the test of a main effect of factor J in a completely randomized experiment as a test of the following H_0: the measurement of each subject (or other experimental unit) is independent of the level of J to which the subject is assigned. This H_0 may be used in reference to all subjects assigned to any of the treatments (cells) or it may be restricted to subjects within certain cells. For instance, given an A × B factorial design, we could test two different H_0's for a main effect of factor A. One H_0 could be this: the measurement of each subject within level B_1 of factor B is independent of the level of A to which the subject is assigned. Alternatively, we could test the following H_0 regarding a main effect of factor A: the measurement of each subject, for any level of B, is independent of the level of A to which the subject is assigned.

In the previous chapters, the test statistics described for randomization tests were equivalent to conventional F and t statistics, so that the computed probability values were the same as they would have been if conventional test statistics had been incorporated into the randomization tests. For randomization tests for factorial designs, however, there will be no attempt to use conventional test statistics or simplified equivalent test statistics because frequently there is disagreement about the appropriate test statistic to use in testing for a certain effect. For example, in a situation where one person uses a test statistic with the "error variance" based on within-cell variability, another person might use the variability associated with interaction as the error variance. Factorial designs, as a result of their complexity, permit the use of a number of alternative test statistics that might be of interest to experimenters, and so the test statistics to be described in this chapter will be chosen for their apparent appropriateness, without consideration of the existence of an equivalent parametric test statistic.

6.3 TEST STATISTICS FOR DESIGNS WITH PROPORTIONAL CELL FREQUENCIES

The test statistics for factorial designs that are appropriate for proportional cell frequencies are different from those that are appropriate for disproportional cell frequency designs, and the two types of designs will be discussed separately. Therefore, it is important to understand what is meant by proportional cell frequencies in a factorial design.

A two-factor design can be represented by a matrix of a × b cells, where there are a levels of factor A and b levels of factor B. The rows of the matrix represent the different levels of one of the factors, and the columns represent the levels of the other factor. Each cell in the matrix represents a distinctive combination of A and B treatments.

In a two-factor experiment, the cell frequencies are said to be completely proportional if the sample-size ratio of any two cells within one row is the same as the sample-size ratio for the corresponding cells in any other row. The matrix in Table 2 is an example of completely proportional cell frequencies. It will be noted that the sample-size ratio of any two cells within a column is the same as the ratio for the corresponding two cells in any other column, although proportionality was defined in terms of

TABLE 2   An Example of Completely
Proportional Cell Frequencies

	a_1	a_2	a_3
b_1	5	5	10
b_2	4	4	8
b_3	7	7	14

cell frequency ratios within rows. Proportionality for rows automatically determines proportionality for columns.

Within a two-factor design there can be proportionality for
a portion of the AB matrix without complete proportionality. For
example, if the frequency for the lower right cell in the matrix
in Table 2 was 10, no longer would there be a completely proportional matrix but there would be proportionality for that portion of the matrix consisting of the first two columns or, alternatively, the first two rows. In such a case, randomization tests
concerning hypotheses including the lower right cell should use
the test statistics to be described later, which are appropriate
for disproportional cell frequencies, whereas other tests could
employ the test statistics that follow in this section, which are
appropriate for proportional cell frequencies.

For a three-factor experiment, cell frequencies for factors A,
B, and C are completely proportional if the AB matrix for each
level of C is proportional, the AC matrix is proportional for each
level of B, and the BC matrix is proportional for each level of A.
With three-factor experiments, as with two-factor experiments,
there can be proportionality for parts of the complete matrix
without complete proportionality.

The simplified test statistics used earlier with single-factor
(one-way) designs also will be used here with multifactor designs
with proportional cell frequencies: $\Sigma(T^2/n)$ for a nondirectional
test of the difference among treatment levels and T_L, the total
for the level predicted to provide the larger mean, for a one-
tailed test. It must be stressed, however, that for most of the
tests to be carried out on factorial designs, the one-way ANOVA
and t test programs are not applicable, despite identity of test

statistics since the data must be permuted differently to generate the distribution of test statistic values.

6.4 PROGRAM FOR TESTS OF MAIN EFFECTS

Program 6.1 can test the effect of any manipulated factor in a randomized design with two or more factors. For a test of the difference between two levels of a factor, the program provides both a one-tailed and a two-tailed P-value. Only one factor at a time is tested, and the input must be organized separately for each test. No matter how many factors are involved in the design, the different cells (combinations of treatments) always can be represented in a two-dimensional array, consisting of rc cells, where r is the number of rows and c the number of columns. If we designate the factor to be tested as A, there are c columns standing for the c levels of A and r rows standing for the number of levels of B, the other factor, for a two-factor design. For designs with more than two factors, the rows stand for combinations of the other factors. Thus, for a three-factor design, with c levels of A, m levels of B, and n levels of C, the complete layout of cells can be represented in this manner:

$$A_1B_1C_1 \qquad A_2B_1C_1 \qquad \cdots \qquad A_cB_1C_1$$

$$A_1B_1C_2 \qquad A_2B_1C_2 \qquad \cdots \qquad A_cB_1C_2$$

$$\cdots \qquad\quad \cdots \qquad\quad \cdots \qquad \cdots$$

$$A_1B_mC_n \qquad A_2B_mC_n \qquad \cdots \qquad A_cB_mC_n$$

where m is the number of levels of B and n is the number of levels of C. There are c columns and mn rows. This is a complete table of all cells, listed in sequential order where, within each column, the subscripts of B and C increment as two-digit numbers. For example, for the first column, after $A_1B_1C_n$, the next cell listed would be $A_1B_2C_1$, which would be followed by $A_1B_2C_2$, and so on.

For a test of the effect of A over all levels of B and C combined, the entire table would be entered into the computer, but for other tests a portion of it would be abstracted and entered. For example, for a test of the difference between A_1 and A_2

PROGRAM 6.1 Test of Main Effects:
Random Permutation

```
      DIMENSION DATA(2000),NOBS(10,10),COLSUM(10),TESTSUM(10)
      DOUBLE PRECISION OBTAIN,SUM,TESTSUM,TEST,COLSUM
      READ(5,)NROWS,NGRPS,NPERM
      NUM=0
      DO 1 I=1,NROWS
      DO 1 J=1,NGRPS
      READ(5,)NOBS(I,J)
    1 NUM=NUM+NOBS(I,J)
      DO 2 I=1,NUM
    2 READ(5,)DATA(I)
      NBACK=0
      DO 3 I=1,NROWS
      DO 3 J=1,NGRPS
      NFRONT=NBACK+1
      NBACK=NBACK+NOBS(I,J)
      SUM=0.
      DO 4 JJ=NFRONT,NBACK
    4 SUM=SUM+DATA(JJ)
    3 COLSUM(J)=COLSUM(J)+SUM
      OBTAIN=0.
      DO 5 J=1,NGRPS
      NSUB=0
      DO 6 I=1,NROWS
    6 NSUB=NOBS(I,J)+NSUB
    5 OBTAIN=OBTAIN+COLSUM(J)**2/NSUB
      IF(NGRPS.EQ.2)OBTAIN1=COLSUM(1)
      PROB=1.
      PROB1=1.
      DO 7 NP=2,NPERM
      TEST=0.
      NX=0
      NBACK=0
      DO 9 I=1,NROWS
      DO 8 IG=1,NGRPS
    8 NX=NX+NOBS(I,IG)
      DO 9 NG=1,NGRPS
      NFRONT=NBACK+1
      NBACK=NBACK+NOBS(I,NG)
      SUM=0.
      DO 10 J=NFRONT,NBACK
      CALL RANDOM_$UNIFORM (R)
      KK=J+R*(NX-J+1)
      X=DATA(KK)
      DATA(KK)=DATA(J)
      DATA(J)=X
   10 SUM=SUM+X
    9 TESTSUM(NG)=TESTSUM(NG)+SUM
      IF(NGRPS.EQ.2)TEST1=TESTSUM(1)
      DO 11 J=1,NGRPS
      NSUB=0
      DO 12 I=1,NROWS
   12 NSUB=NOBS(I,J)+NSUB
      TEST=TEST+TESTSUM(J)**2/NSUB
   11 TESTSUM(J)=0.
      IF(NGRPS.EQ.2.AND.TEST1.GE.OBTAIN1)PROB1=PROB1+1.
    7 IF(TEST.GE.OBTAIN)PROB=PROB+1
      PROB=PROB/NPERM
      PROB1=PROB1/NPERM
      WRITE(6,50)PROB
      IF(NGRPS.EQ.2)WRITE(6,60)PROB1
   50 FORMAT(///10X,"PROBABILITY FOR ANOVA IS ",F8.6)
   60 FORMAT(///10X,"ONE-TAILED PROBABILITY IS ",F8.6)
```

over all levels of B and C, the first two columns would be used, whereas if H_0 concerned differences among all levels of A within all levels of C and level 1 of B, the first n rows would be fed into the computer. When only two levels of A are to be compared, the level of A designated as A_1, should be the level predicted to provide the larger measurements, because the one-tailed test statistic T_L is the total of the first column.

After determining the portion of the table to be analyzed, the following steps are performed in applying Program 6.1:

Step 1. The user must specify the number of rows (NROWS) and the number of columns (NGRPS) in the array and the number of measurements in each cell, cell by cell, starting with the first cell in the first row, then the second cell in the first row, and so on, ending with the last column in the last row. The data are entered in the same sequence.

Step 2. NPERM, the number of permutations upon which to base the test, is specified. One of the data permutations is that associated with the experimental results and the other NPERM − 1 data permutations are randomly generated.

Step 3. If only two levels of a factor are being tested, the computer performs step 6 for the obtained data to compute the obtained test statistic value.

Step 4. The computer performs steps 7 and 8 for the obtained data to compute $\Sigma(T^2/n)$, the nondirectional test statistic value.

Step 5. The computer uses a random-number generation procedure to randomly divide the data in the first row among the columns, holding cell frequencies fixed, and does the same for each successive row. A division of the data carried out on all rows is defined as a single permutation. NPERM − 1 permutations are performed and for each permutation steps 6 to 8 are carried out.

Step 6. The computer computes T_L, the total of the measurements for the first column of the array. NOTE: The level of A selected to be the first column should be the level predicted to provide the larger measurements. Step 6 is carried out only when there are just two levels of A being tested.

Step 7. The computer computes T_1, . . . , T_c, the totals for each of the columns.

Step 8. The computer squares each T and divides each T^2 by n, the number of measurements in the column, then adds the values of T^2/n to get $\Sigma(T^2/n)$, the nondirectional test statistic.

Step 9. The computer computes the P-values for T_L (when appropriate) and $\Sigma(T^2/n)$ as the proportion of the NPERM test statistic values that are greater than or equal to the obtained values.

Program 6.1 is a general-purpose program. In addition to performing special tests for factorial designs, it can perform independent and correlated t tests and both one-way and repeated-measures ANOVA.

6.5 COMPLETELY RANDOMIZED TWO-FACTOR EXPERIMENTS

Sometimes experimenters want to investigate the effects of two experimentally manipulated variables (factors) in a single experiment. An example will show some of the randomization tests that can be performed on data from such experiments.

Example 6.1

A botanist conducts a greenhouse experiment to study the effects of both temperature and fertilizer on plant growth. He uses two levels of temperature, called High and Low, and two levels (types) of fertilizer, called R and S. Twenty plants are randomly assigned to the four treatment cells (combinations of temperature and fertilizer levels), with the sample size constraints shown in Table 3. Table 4 presents the results of the experiment. First, a test is conducted to test the following H_0: the growth of each plant is independent of the type

TABLE 3   Number of Plants Assigned to Each Experimental Condition

		Fertilizer Type	
		R	S
Temperature	High	4	4
	Low	6	6

TABLE 4   Experimental Results

	Fertilizer Type	
	R	S
High temperature	12	15
	11	18
	14	17
	15	18
	$\overline{X} = 13$	$\overline{X} = 17$
Low temperature	16	22
	17	20
	14	17
	15	20
	18	21
	16	20
	$\overline{X} = 16$	$\overline{X} = 20$
	$\overline{X} = 14.8$	$\overline{X} = 18.8$

of fertilizer applied to it. In testing this H_0, we want to be sure that the test is valid whether or not there is a temperature effect. Consequently, for testing H_0 the only relevant data permutations are those where data are permuted within levels of temperature but not between temperature levels. Since H_0 does not refer to temperature effects, other permutations are not relevant. The significance is determined by reference to that subset of data permutations where the data for the two temperature levels, shown in Table 4, are permuted within those two levels. There are $(8!/4!4!) \times (12!/6!6!) = 64,680$ data permutations, since each of the $(8!/4!4!)$ divisions of the measurements for high temperature between the two levels of fertilizer can be associated with each of the $(12!/6!6!)$ divisions of the measurements for low temperature between the two types of fertilizer. As there are 64,680 data permutations in the subset, the experimenter decides to use a random data permutation procedure, instead of a systematic procedure, to determine significance.

Program 6.1 is used by the experimenter to determine significance of the effect of fertilizer. For each random data permutation, the program randomly divides the measurements within both levels of temperature between the fertilizer types and computes the appropriate test statistics for those data permutations.

For the obtained results, the value of the two-tailed test statistic, $\Sigma(T^2/n)$, for the effect of fertilizer is computed in this way: $\Sigma(T^2/n) = (148)^2/10 + (188)^2/10 = 5724.8$. In other words, the obtained value of $\Sigma(T^2/n)$ is the R total squared divided by the number of R measurements plus the S total squared divided by the number of S measurements. For each data permutation $\Sigma(T^2/n)$ is computed, and the proportion of data permutations with $\Sigma(T^2/n)$ as large as 5724.8 is the significance value for the fertilizer effect.

For a one-tailed test where type S fertilizer was predicted to provide the larger mean, the test statistic used would be the S total, which for the obtained data is 188.

Inasmuch as the plants were assigned randomly to levels of both factors, the data can be permuted to determine the main effect for either factor. In addition to finding the significance of the main effect of fertilizer, one also can find the significance of the main effect of temperature. The null hypothesis for testing the main effect of temperature is: the growth of each plant is independent of the level of temperature. For this test the only relevant data permutations are those associated with assignments to levels of fertilizer actually made in the experiment. Significance is based on the subset of data permutations where the data for the two fertilizer types are permuted within those levels. There are $(10!/4!6!) \times (10!/4!6!) = 44,100$ data permutations, since each of the divisions of the measurements for type R fertilizer between the two levels of temperature can be associated with each of the divisions of the measurements for type S fertilizer between the two levels of temperature.

Again, we will consider significance determination by random data permutation, using Program 6.1. For each data permutation there is computation of $\Sigma(T^2/n)$ where T is a total for a level of temperature and n is the number of plants at that level. Thus, $\Sigma(T^2/n)$ for the obtained results = $(120)^2/8 + (216)^2/12 = 5688$. The proportion of the data permutations produced by Program 6.1 with $\Sigma(T^2/n)$ as large as 5688, the obtained value, is the P-value associated with the main effect of temperature.

The alternative hypothesis that would be accepted if the H_0 was rejected is this: some of the plants would have grown a different amount if they had been assigned to a different level of temperature.

For a one-tailed test, the total for the temperature level predicted to provide the larger mean is used as the test statistic in Program 6.1, and the data for those cells should be entered into the computer before the data for the other cells.

Suppose there was interest in testing the effect of fertilizer within the high level of the temperature variable. The null hypothesis to test is that the growth of each of the eight plants assigned to the high level of temperature is independent of assignment to level of fertilizer. The relevant data permutations for testing H_0 are those associated with assignment of the same eight plants to the high level of temperature. The top part of the data matrix in Table 4 is abstracted and analyzed as a separate unit. The systematic data-permuting program for the independent t test, Program 4.4, can be applied to determine one-tailed and two-tailed P-values, as there are only $8!/4!4! = 70$ data permutations. Rejection of H_0 implies acceptance of the following H_1: some of the eight plants assigned to the high level of temperature would have grown a different amount if they had been assigned to the other type of fertilizer.

Next, let us consider a completely randomized two-factor experiment in which one factor is introduced solely for the purpose of providing a more sensitive test of the other factor. In Example 2.2, there was a description of random assignment in which subject-plus-treatment-time-plus-nurse was a systematically constructed experimental unit which was randomly assigned to a drug or placebo condition. If the nurses actually influence the results, the two-factor experiment described in the following example might provide a more sensitive test.

Example 6.2

In designing an experiment to compare the effects of a drug and a placebo, a medical researcher finds it necessary to employ two different nurses to administer the treatments. It would have been preferable to control for the nurse variable by using only one nurse, but as that was not feasible, the researcher introduces the nurse as a factor in a two-factor completely randomized design. The design and the number of subjects assigned to each condition are shown in Table 5.

TABLE 5   Number of Subjects for Each
Treatment Condition for Nurses A and B

	Drug	Placebo
Nurse A	10	10
Nurse B	10	10

Each of the 40 subjects used was allowed to specify a time for
receiving a treatment, and the subject-plus-treatment-time
experimental units were randomly assigned to the four cells of
the matrix, with 10 subjects per cell.

The medical researcher is not interested in testing the effect
of the nurses; the only main effect of interest is the drug
effect. For this test, the 20 measurements within each row
would be permuted randomly over the drug and placebo conditions and the one- or two-tailed test statistic (i.e., T_L or
$\Sigma(T^2/n)$) would be computed for those data permutations for
determining significance by means of a randomization test. By
permuting the data only within rows, the variability due to
nurse effect is controlled, leading to a more sensitive test than
could be performed on data from the assignment in Example 2.2
by applying a t test.

6.6   COMPLETELY RANDOMIZED THREE-FACTOR
EXPERIMENTS

Now we will consider an extension of the completely randomized
factorial design to more than two factors. An extension of randomization tests to three factors will generalize to completely
randomized designs with more than three factors.

Example 6.3

An experiment is conducted where all three factors are experimentally manipulated. Factor A has two levels, factor B has
three levels, and factor C has two levels. The various possible
treatments are represented as cells in the array shown in Table
6. The random assignment of subjects to cells is performed
within the constraint imposed by predetermined cell frequencies.
Given 60 subjects to be assigned, five per cell, there would be

TABLE 6   Three-factor Experimental Design

	A_1		A_2	
	C_1	C_2	C_1	C_2
B_1	$n = 5$	$n = 5$	$n = 5$	$n = 5$
B_2	$n = 5$	$n = 5$	$n = 5$	$n = 5$
B_3	$n = 5$	$n = 5$	$n = 5$	$n = 5$

$60!/(5!)^{12}$ possible assignments. The effect of factor A within level 1 of factor B and level 1 of factor C can be tested by computing F for one-way ANOVA for each of the $10!/5!5! = 252$ data permutations where the 10 obtained measurements in the first and third cells of the top row are permuted. Restriction of the analysis to one level of each of two factors reduces the three-factor design to a one-factor design, and so the data permutations and the F for main effects are those for one-way ANOVA.

The main effect of factor A within level 1 of factor B, over both levels of factor C, also can be tested. For a randomization test based on random data permutation, $\Sigma(T^2/n)$ and T_L are computed for a random sample of the $(10!/5!5!)^2$ data permutations where the 10 measurements obtained in the first and third cells of the top row are divided between those two cells in every possible way and, for each of those divisions, the 10 obtained measurements for the second and fourth cells in the top row are interchanged in every possible way. Program 6.1 could be used to determine the P-values for the two test statistics.

The effect of factor A over all levels of the other two factors can be tested also. The number of data permutations involved would be the number of ways the data can be divided between levels 1 and 2 of factor A without moving any measurement to a level of factor B or factor C different from the level in which it actually occurred in the experiment. Thus data transfers could take place only between the first and third cells and between the second and fourth cells within each row. Consequently, the number of data permutations would be $(10!/5!5!)^6$. Significance can be determined by use of Program 6.1, based on a random sample of the $(10!/5!5!)^6$ data permutations.

6.7   INTERACTIONS IN COMPLETELY RANDOMIZED EXPERIMENTS

In the following discussion of interactions in 2×2 factorial designs, the upper left cell is cell 1, the upper right is cell 2, the lower left is cell 3, and the lower right is cell 4.

The null hypothesis of no interaction for a 2×2 design, for the conventional random sampling model, can be expressed in terms of population means in the following way: $\mu_1 - \mu_2 = \mu_3 - \mu_4$, where the subscripts denote the different cells. For the typical experiment, in which there is no random sampling, a null hypothesis expressed in terms of population parameters is not relevant. However, we can formulate an analogous H_0 of no interaction, expressed in terms of a sampling distribution, without reference to population parameters: $E(\bar{X}_1 - \bar{X}_2) = E(\bar{X}_3 - \bar{X}_4)$, where E refers to the expectation or average value of the expression following it, the average being based on all possible assignments. Unfortunately, this H_0 does not refer to measurements for individual subjects, and so there is no basis for generating data permutations for the alternative subject assignments, a necessity for determining significance by data permutation.

Let us define the H_0 of no interaction in a way that can be used to associate data permutations with alternative subject assignments: for every subject, $X_1 - X_2 = X_3 - X_4$. The null hypothesis thus refers to four measurements for each subject. For a 2×2 completely randomized design, where each subject takes only one of the four treatments, one of the four X's represents the subject's actual measurement for that treatment and the other three X's refer to potential measurements which the subject would have provided under alternative assignments.

We will now make explicit some aspects of the H_0 of no interaction. To do this, first we will expand the statement of H_0: for every subject, $X_1 - X_2 = X_3 - X_4 = \theta$, where θ may be 0 or any positive or negative value for a subject and where the value of θ may vary in any way whatsoever over subjects. Thus H_0 would be true if, for every subject, $X_1 - X_2 = X_3 - X_4 = 0$; but it would be true also if, for some subjects, the value of θ was 0, for others it was -4, and for others it was $+8$.

Example 6.4

Let us now examine a test of this H_0 in a completely randomized experiment, where the following measurements were obtained from eight subjects:

	A_1	A_2
B_1	3, 4	1, 2
B_2	7, 8	5, 6

Inasmuch as $\overline{X}_1 - \overline{X}_2 = \overline{X}_3 - \overline{X}_4$ for these results, the computation of F for interaction would give an interaction F with a value of 0. Using F for interaction as a test statistic, the obtained test statistic value is thus 0. Now let us consider a data permutation associated with the random assignment where the subject receiving the 7 was assigned to cell 4 and the subject receiving the 5 was assigned to cell 3. Under H_0, the data permutation associated with such a random assignment can be represented as follows:

	A_1	A_2
B_1	3, 4	1, 2
B_2	$5 + \theta_1$, 8	$7 - \theta_2$, 6

Under H_0, θ_1 and θ_2 can be, but need not be, equal. Suppose that θ_1 and θ_2 both have values of 0. Then, for the above data permutation, we would have a value of interaction F greater than 0, because $\overline{X}_1 - \overline{X}_2$ would not equal $\overline{X}_3 - \overline{X}_4$. On the other hand, suppose that θ_1 and θ_2 both equal 2. Then the above data permutation would be the same as the obtained results, and would have an interaction F of 0. Thus, the value of interaction F associated with the above data permutation is indeterminate, and that is true of any data permutation except the one associated with the obtained results. Consequently, since the proportion of data permutations with as large a value of interaction F as the obtained value is a function of the value of θ, the H_0 for interaction cannot be tested, because it does not specify the value of θ. Even for the very simplest of factorial designs for independent groups, therefore, there is no way to test the H_0 of no interaction by data permutation, and the problem remains as more factors are added to the design.

Even though there is no obvious meaningful H_0 of no interaction for a completely randomized design that can be tested by a randomization test procedure, one could use F for interaction of A and B as a test statistic to test the overall H_0 of no effect of either factor. To test the H_0 of no treatment effect by use of F for interaction as a test statistic, the data can be permuted over all cells in the AB matrix, with the interaction test statistic computed for each data permutation for a randomization test based on systematic permutation. The total number of data permutations can be readily determined. The number of data permutations would equal the number of possible assignments in the completely randomized design. For random data permutation, the data would be permuted randomly throughout the matrix. As this is a test of no effect of either factor, it is not necessary to permute only within rows or only within columns. For each data permutation, F for interaction would be computed as the test statistic to provide the reference set for determining the significance of the obtained interaction F.

Tests that have been described in articles as "randomization tests for interaction" for completely randomized factorial designs thus may be useful as tests of no treatment effect that are sensitive to interaction effects. But the proposed randomization tests frequently are faulty as a result of trying to provide a simpler computation of the test statistic than that required for interaction F for each data permutation. For instance, it may be proposed that the obtained AB data matrix be transformed to adjust the measurements for A and B main effects. Each measurement is transformed into a residual value showing its deviation from the row and column means, and an interaction test statistic is computed from the residuals. Unfortunately, after computing the obtained test statistic from the residuals, the residuals are then permuted to generate the other data permutations from which test statistics are computed, instead of permuting the raw data and generating a new set of residuals for each data permutation. Permuting the residuals does not provide a set of residuals that under the H_0 of no treatment effect would have been obtained under alternative assignments and in fact may provide matrices of residuals which would have been impossible under the H_0, given the obtained data.

Instead of performing a randomization test of the null hypothesis of no treatment effect, using a conventional interaction test statistic, a researcher may find it useful to test the H_0 of no effect of a particular factor, using a test statistic sensitive to disordinal interactions, as when the effect of factor A at one level of B is opposite to its effect at the other level of B in a 2 × 2 factorial design. Now, if we had expected the direction of effect of A to be the same for both levels of B, we would have permuted the data within levels of B. When the

direction of effect is expected to be opposite we still permute
data only within levels of B. What must be changed in the test
is not the method of permuting data but the computation of a
test statistic that will reflect the different prediction. Consider
the following 2 × 2 factorial design:

	A_1	A_2
B_1	cell 1	cell 2
B_2	cell 3	cell 4

If we expected level A_1 to provide larger values within level B_1
and level A_2 to provide larger values within level B_2, we could
use the combined total of the scores in cells 1 and 4 (T_L) as
our one-tailed test statistic that would be computed over permutations of data between cells 1 and 2 and between cells 3 and 4.
For a two-tailed test, where the direction of difference between
means was expected to be opposite within the two levels of B
but where there was no basis for specifying which two cells
should have the larger means within their row, the data would
be permuted in the same way as for the one-tailed test, but
the test statistic would be $\Sigma(T^2/n)$ instead of T_L. $\Sigma(T^2/n)$
would be (combined total of cells 2 and 3)^2/(total number of
measurements in cells 2 and 3) + (combined total of cells 1 and
4)^2/(total number of measurements in cells 1 and 4).

The data can be arranged differently prior to entry into
the computer when a disordinal interaction is predicted in order
to see more readily what combinations of cells are to be compared.
For example, we could arrange the AB matrix in this way,
reversing the order of the two cells in the second row:

A_1B_1	A_2B_1
A_2B_2	A_1B_2

If the data were entered into the computer in that form, Program
6.1 would perform the one-tailed test described above by using
the total of the measurements in the left column as T_L, and the
two-tailed test with the test statistic $\Sigma(T^2 n)$, would be based

on the column totals and the total frequency of measurements
within the columns. This arrangement makes it clear that we
expect large measurements when the levels of A and B are the
same and small ones otherwise. A psychological experiment
where such a prediction might be appropriate would be one in
which the two levels of B were amount of work required of a
subject and the levels of A were amount of pay given for the
work, with the measurements being the calmness in accepting
the pay. The right column in the matrix above would represent
responses of subjects who might be surprised and excited at
the discrepancy between the amount of work performed and their
pay for it.

In some factorial designs there is no low or high level for
a factor, as in Example 6.1, where the levels for the fertilizer
factor were simply types R and S. Expectation of reversal of
direction of effect of fertilizers within low and high levels of
temperature could lead to a randomization test for column differences where the cells in one of the rows were switched, but
that expectation would not be based on notions of homogeneity
of levels being important, as in the psychological experiment
just given because of the arbitrariness in designating type R or
S fertilizer as level 1 or 2. There could, however, be alternative reasons for expecting a disordinal interaction, such as the
effect of temperature on the chemicals in the two types of
fertilizer.

6.8   RANDOMIZED BLOCK EXPERIMENTS

In factorial experiments, the factors need not all be ones that
are experimentally manipulated. For instance, it is conventional
to designate an experiment where one factor is experimentally
manipulated and another factor is a stratifying or blocking
variable, like kind of subject, as a two-factor experiment. When
the subjects have not been selected randomly but are divided
systematically into blocks according to age, sex, or some other
property of the subjects, the only statistical tests of main effects
that can be conducted by data permutation are those that concern the main effect of the manipulated variable. No randomization test can be performed to determine the main effect of a
subject variable, because subjects cannot be assigned randomly
to different levels of a subject variable. Stratifying (blocking)
subjects according to a subject variable, however, can be useful
in increasing the probability of detecting an effect of the manipulated variable.

From the standpoint of randomization tests, the purpose of the type of randomized block design known as a treatments-by-subjects design is similar to that of repeated-measures designs: to control for between-subject variability and thereby provide a more sensitive test of treatment effects. If different kinds of subjects tend to provide quite different magnitudes of response to an experimentally manipulated variable, the variability within the total group of subjects may be quite large relative to the difference between treatment means. On the other hand, within each kind (level) of subject the variability may be small. When that is the case, the use of a treatments-by-subjects design controls for variability over different kinds of subjects and thus provides a more sensitive test of treatment effects.

In Chapter 5, we considered a treatments-by-subjects design in which the number of subjects in each block was the number of treatments, a design called a matched-subjects design. That design was considered in Chapter 5 because it can be analyzed by the repeated-measures computer program. A treatments-by-subjects design with more subjects per block than the number of treatment levels cannot be analyzed by the programs in Chapter 5, but Program 6.1 is applicable.

Example 6.5

An experimenter uses six males and six females to compare the effectiveness of three treatments, A, B, and C. He randomly assigns two males to each treatment and then randomly assigns two females to each treatment. It is a randomized block experiment, because within both the male and female blocks there is random assignment to treatments, but inasmuch as the subjects are not randomly assigned to the subject variable, sex, it is not a completely randomized experiment. Table 7 shows the results of the experiment.

First, the experimenter tests this H_0: the measurement of each subject is independent of the level of treatment. A test is carried out to test the main effect of the treatment variable. $\Sigma(T^2/n)$ is a test statistic that can be used for this purpose. The value of $\Sigma(T^2/n)$ for the obtained results shown in Table 7 is $(24)^2/4 + (32)^2/4 + (52)^2/4 = 1,076$. For a systematic randomization test, $\Sigma(T^2/n)$ is computed for each of the $6!/2!2!2! \times 6!/2!2!2! = 8,100$ data permutations, in which the six measurements within each level of the sex variable are divided among the three treatments. The proportion of the 8,100 data permutations providing as large a value of $\Sigma(T^2/n)$

TABLE 7   Treatments-by-subjects Design

	A	B	C	
Males	4	6	8	
	6	8	10	$\overline{X} = 7$
	$\overline{X} = 5$	$\overline{X} = 7$	$\overline{X} = 9$	
Females	6	8	17	
	8	10	17	$\overline{X} = 11$
	$\overline{X} = 7$	$\overline{X} = 9$	$\overline{X} = 17$	
	$\overline{X} = 6$	$\overline{X} = 8$	$\overline{X} = 13$	

as 1,076 is the P-value for the main effect of the treatment variable. Program 6.1 can be used to perform this test by use of random data permutation. Rejection of H_0 implies acceptance of the following H_1: some of the subjects would have given different measurements under alternative treatments.

The main effect of treatments on the males or females alone can be tested by extracting either the top or the bottom half of the table and computing $\Sigma(T^2/n)$ for that portion of the table for all $6!/2!2!2! = 90$ data permutations. The null hypothesis is that the measurement for each of the six persons of that sex is independent of the level of treatment. Significance can be determined by using a program from Chapter 4, because $\Sigma(T^2/n)$ for the males or females alone is $\Sigma(T^2/n)$ for one-way ANOVA, and the data permutation procedure also is that of one-way ANOVA. Rejection of H_0 implies acceptance of the following H_1: the measurements of some of the six persons of that sex would have been different under alternative treatments.

If desired, one- or two-tailed tests could be conducted on pairs of treatments by extracting the two relevant columns and using $\Sigma(T^2/n)$ or T_L as the test statistic. For a systematic test, the data would be permuted in $4!/2!2! \times 4!/2!2! = 36$ ways, dividing the data repeatedly within male and female levels. A systematic program would be appropriate for this test, but the random permuting program, 6.1, is quite valid for this application.

Example 6.6

A three-factor experiment, in which two factors are manipulated and one factor is a subject variable, is conducted. Factor A has levels A_1 and A_2, and factor B has levels B_1 and B_2. The third factor is a blocking variable, sex, with two levels, Male and Female. Each combination of levels of the A and B factors represents a distinctive treatment. Two males are assigned randomly to each of the four distinctive treatments, and two females are assigned randomly to each of the treatments. Thus there are $8!/2!2!2!2! \times 8!/2!2!2!2! = 6,350,400$ possible assignments. The results of the experiment are shown in Table 8.

A test of the main effect of factor A over both levels of factor B and both levels of the subject variable would be conducted in the following way. For a one-tailed test, the test statistic T_L is computed for the obtained results. If the first level of factor A had been expected to provide the larger measurements, the value of T_L would be 68. For a two-tailed test, we compute $\Sigma(T^2/n)$ for the main effect of factor A for the obtained results, which is $(68)^2/8 + (88)^2/8 = 1,546$. To determine significance by data permutation, the data are permuted according to the following H_0: the measurement for each subject is independent of assignment to level of factor A. Inasmuch as H_0 does not refer to the effect of factor B or the effect of factor C, the subject variable, data must be divided between

TABLE 8   Three-factor Experimental Design

	A_1		A_2	
	Males	Females	Males	Females
B_1	4	7	8	10
	7	8	9	11
	$\bar{X} = 5.5$	$\bar{X} = 7.5$	$\bar{X} = 8.5$	$\bar{X} = 10.5$
B_2	9	10	11	12
	12	11	13	14
	$\bar{X} = 10.5$	$\bar{X} = 10.5$	$\bar{X} = 12$	$\bar{X} = 13$
		$\bar{X} = 8.5$		$\bar{X} = 11$

cells that are at the same level of variable B and at the same
level of C, the subject variable. (Of course, we could not permute data over levels of C, the subject variable, even to test
its effect, because of lack of random assignment to levels of
that variable.) Designating the top row of cells in Table 8
as cells 1 to 4 and the bottom row as cells 5 to 8, we use the
data permutations consisting of all possible combinations of divisions of data within the following pairs of cells: 1 and 3,
2 and 4; 5 and 7; 6 and 8. Since there are $4!/2!2!$ data divisions for each of these pairs of cells and since every data division for a pair of cells can be associated with every data division for each of the other pairs of cells, there are $(4!/2!2!)^4 =$
1,296 data permutations. The proportion of data permutations
with as large a value of T_L as 68, or as large a value of
$\Sigma(T^2/n)$ as 1,546, as determined by Program 6.1, is the P-value
for the test of the main effect of A.

   To test the effect of factor A within level B_1, we extract
the first row of data from Table 8 and compute $\Sigma(T^2/n)$ for
the obtained results as $(26)^2/4 + (38)^2/4 = 530$. T_L for the
obtained results would be 26, for a prediction that the first
level of A would provide the larger measurements. The eight
measurements within level B_1 are permuted in $4!/2!2! \times 4!/2!2! =$
36 ways, consisting of every division of the four measurements
for males between the cells for males in conjunction with every
division of the four measurements for females between the cells
for females. The significance can be determined by use of
Program 6.1.

6.9   RANDOMIZED BLOCK EXPERIMENTS WITH
REPEATED MEASURES

Repeated-measures experiments, in which each subject takes all
treatments at treatment times assigned randomly to treatments
independently for each subject are regarded by some as randomized block experiments where individual subjects, rather than
kinds of subjects, constitute blocks. There are, nevertheless,
fundamental differences between repeated-measures experiments
and experiments in which kinds of subjects constitute blocks.
In the first place, treatments-by-subjects experiments may be
possible when repeated-measures experiments are not. For
example, the same seed cannot be planted in two different kinds
of soil at the same time, nor can an animal be raised from birth
to 6 weeks of age under total darkness and concurrently under

normal illumination conditions. Secondly, carryover effects within subjects in repeated-measures experiments create problems of interpretation of results that can be avoided easily in treatments-by-subjects designs.

Let us consider an example of a two-factor experiment in which each subject is assigned to a level of one factor and each subject takes all levels of the other factor. Such designs may occur because of the desirability (and feasibility) of repeated measures for some factors and not for others.

Example 6.7

An experiment with three levels of factor A and two levels of factor B is conducted. Each subject is assigned to all three levels of factor A and to one of the two levels of factor B. The four subjects are assigned in the following manner. Each subject has three designated treatment times and for each subject, independently, there is random assignment of the three treatment times to levels 1, 2, and 3 of factor A. Then there is random assignment of two of the subjects to level 1 of factor B and the remaining two subjects are assigned to level 2. Table 9 shows the results of the experiment, where A and B are the factors and S's are the subjects.

H_0 for the main effect of factor A is: at each of the three treatment times the measurement of a subject is independent of the level of A. There are $3! = 6$ ways in which the three measurements for each subject can be divided among the three levels of A, and each of those divisions can be associated with six

TABLE 9   Factorial Design with Repeated Measures

		A_1	A_2	A_3
B_1	S_1	5	7	9
	S_2	7	11	12
B_2	S_3	3	5	7
	S_4	5	9	10

divisions for each of the other subjects, making a total of $(3!)^4 =$
1.296 data permutations. The generation of data permutations
is the same as for univariate repeated-measures ANOVA, and the
test statistic, ΣT^2, reflects the property of interest in assessing the effect of A, namely, the variation of the A means over
levels of A. ΣT^2 for the obtained data, shown in Table 9, is
$(20)^2 + (32)^2 + (38)^2 = 2868$. The value of ΣT^2 is computed
for each of the 1,296 data permutations, and the proportion of
the 1,296 test statistic values that are as large as the obtained
value is the P-value for the main effect of factor A. Program
5.1 can be used to determine the P-value.

H_0 for the main effect of factor B is: the measurement of
each subject at each of the levels of A is independent of the
subject's assignment to level of B. For example, if H_0 is true
the row of measurements for S_1 would have been a row for level
B_2 if S_1 had been assigned to level B_2. One hypothesis that
can be tested is that of no effect of B at either level of A. For
the test of the null hypothesis, the three measurements for each
subject can be added to provide a single sum. Then the four
sums are divided in all $4!/2!2! = 6$ ways between the two levels
of B, and the one-tailed and two-tailed test statistics, T_L and
$\Sigma(T^2/n)$, are used for a randomization test. A computer program is unnecessary for this simple example, but for larger
sample sizes, programs from Chapter 4 could be employed.
Three other null hypotheses that could be tested would be those
of no effect of B at a particular level of A. For testing each
of those three null hypotheses, the four measurements for that
level of A could be divided between the two levels of B with
two measurements for B_1 and two for B_2 in all possible ways by
means of a program from Chapter 4, to determine significance.

6.10   INTERACTIONS IN REPEATED-MEASURES
EXPERIMENTS

In Section 6.7, a randomization test counterpart of the H_0 of no
interaction for a population model was described, and it was
shown that it could not be tested for a completely randomized
experiment although an interaction test statistic could be used
for testing other H_0's. On the other hand, it *is* possible to
test a H_0 of no interaction for a *repeated-measures* design in
which there are subjects providing repeated measures over levels
of factor A within different levels of factor B. The following

example is based on a previously published discussion (Edgington, 1969, pp. 118–120).

Example 6.8

In a two-factor experiment, factor A and factor B both have two levels. Each of 10 subjects has repeated measures over the two levels of A but is assigned randomly to one of the levels of B. Each subject has two designated treatment times which are assigned randomly to levels A_1 and A_2 independently for each subject. Then the experimenter selects 5 of the 10 subjects at random for assignment to level B_1 and the other five are assigned to level B_2. The obtained measurements for this experiment are given in Table 10.

The null hypothesis of no interaction in the experiment is: for each of the 10 subjects, the difference score $(A_1 - A_2)$

TABLE 10   Data for Test of Interaction

		A_1	A_1	A_2	$A_1 - A_2$
B_1	S_1	8	5	3	
	S_2	7	4	3	
	S_3	9	9	0	
	S_4	5	6	-1	
	S_5	10	7	3	
B_2	S_6	16	10	6	
	S_7	13	9	4	
	S_8	12	14	-2	
	S_9	19	15	4	
	S_{10}	22	17	5	

shown in the last column of Table 10, is independent of assignment of the subject to a level of the B factor. There are $10!/5!5! = 252$ possible assignments to levels of B and, under H_0, the difference score for a subject is that which that subject would have had under either assignment. Thus, there are 252 data permutations representing all possible divisions of the 10 difference scores into two groups of five each. The test statistic $\Sigma(T^2/n)$, where T is the total of the difference scores for a treatment, can be used to test H_0 for a two-tailed test. For the obtained results the test statistic value is $8^2/5 + 17^2/5 = 70.6$. The proportion of the data permutations with as large a value of the test statistic as 70.6 is the P-value associated with the interaction. The P-value can be determined by using a one-way ANOVA program, in Chapter 4, with the 10 difference scores constituting the data to be analyzed. Rejection of H_0 implies acceptance of the following H_1: for at least one subject the difference score would have been different for the alternative level of B. For a one-tailed test where the difference scores for level B_2 were predicted to be larger, the P-value would be half of that for the two-tailed test because of the equality of number of difference scores in the two levels of B. For situations with unequal sample sizes for B_1 and B_2, the test statistic T_L (that is, the total of B_2) could be used, with significance determined by one of the independent t test programs in Chapter 4.

6.11   DISPROPORTIONAL CELL FREQUENCIES

Earlier it was stated that cell frequencies are completely proportional in a two-factor experiment if the sample size ratio of any two cells in one row is the same as the sample size ratio for the corresponding cells in any other row. When the sample size ratio varies over the rows for corresponding cells, there are said to be disproportional cell frequencies. Table 11 is an example of *disproportional* cell frequencies of the kind that could result from the loss of two subjects from the lower right cell in an experiment that started with proportional frequencies. For analyses involving only the a_1 and a_2 levels of factor A, the cell frequencies are proportional and the same is true for analysis of only the b_1 and b_2 levels of factor B. Consequently, such analyses could be run according to the procedures considered previously, which assume proportional cell frequencies. Any analysis which involves the lower right cell, however,

TABLE 11 Disproportional
Cell Frequencies

	a_1	a_2	a_3
b_1	3	4	5
b_2	6	8	10
b_3	9	12	13

should not employ the standard procedure but either the procedure that will be described in Section 6.12 or the one in Section
6.14.

What will be proposed is the use of an alternative test statistic for designs with disproportional cell frequencies. The test
statistic $\Sigma(T^2/n)$ is valid for disproportional as well as proportional cell frequencies when significance is based on data permutation, but the test statistic to be developed here is more
powerful when cell frequencies are disproportional.

Before considering the new test statistic, let us see why one
might have a factorial design with disproportional cell frequencies. When there are several separate single-factor experiments
run at different times to provide data for a treatments-by-
replications design, the cell frequencies might be disproportional
for a number of reasons. Each experiment might employ whatever subjects are available at the time, and the number of subjects may not be an exact multiple of the number used in another
experiment, making it impossible to maintain proportionality of
cell sizes. In some cases the experimenter may deliberately
introduce disproportionality. For example, if the variation within
one level of a factor is so great that the experimenter feels there
should be more subjects at that level in order to detect a treatment effect, in later experiments, a larger proportion of subjects may be assigned to that level.

Factorial designs based on single experiments can also involve
disproportionality of cell size. Obviously the number of available subjects could at times prevent an experimenter from using
all subjects if he insisted on *equal* sample sizes for all cells.
Furthermore, the number of available subjects might not fit into
some other proportional cell size design. Also, in single experiments as well as in a series of experiments, experiments designed

with proportional cell sizes may become experiments with disproportional cell sizes because of subject mortality or failure of
some assigned subjects to participate in the experiment.

Still another reason for disproportionality of cell frequencies
is the use of a blocking factor in the analysis that was not
planned as a blocking factor in the assignment to treatments.
In Example 6.2 we considered an experiment for determining the
effect of a drug with nurse as a factor, the levels being Nurse
A and Nurse B, in order to have a more sensitive test of drug
effects. Suppose, however, that the nurse factor had not been
planned for the experiment and that the assignment was conducted as in Example 2.2, where each subject was associated
systematically with either Nurse A or Nurse B and then the
subject-plus-nurse experimental units were assigned randomly to
drug or placebo conditions. Although planned as a single-
factor study, the study provides results that could be represented in a two-factor layout with four cells for combinations of
the two treatment conditions and the two nurses. Nominally,
the two nurses are levels of the nurse factor, but the measurements in the two levels may differ systematically because of the
subjects associated with the nurses. That consideration, however, does not complicate the analysis. If it is reasonable to
expect the measurements associated with the subjects injected
by one nurse to differ systematically from those associated with
the subjects injected by the other nurse, then dividing the
measurements within two levels of the nurse factor may provide
a more sensitive test than a single-factor test. Suppose the
assignment procedure randomly divided 20 subject-plus-nurse
units into 10 for placebo and 10 for drug and that each nurse
injected 10 subjects. The results of the study could lead to
disproportionality of cell size when the nurse factor is introduced
into the layout, as shown in the following distribution of cell
frequencies:

	Drug	Placebo
Nurse A	6	4
Nurse B	4	6

Although the assignment did not restrict the cell frequencies as
shown above, for purposes of analysis by a randomization test
procedure, the 10 measurements for Nurse A could be divided
between Drug and Placebo columns in 10!/6!4! ways, in

conjunction with the division of the 10 measurements for Nurse B between Drug and Placebo columns in 10!/4!6! ways. The test statistic to be described in Section 6.12 or the procedure to be described in Section 6.14 would be appropriate. Whether such after-the-fact blocking would provide a more sensitive test when cell frequencies are small is questionable, but it is a valid procedure that might prove useful with large sample sizes.

6.12 TEST STATISTIC FOR DISPROPORTIONAL CELL FREQUENCIES

The following example will show why a special test statistic should be used when cell frequencies are disproportional.

Example 6.9

In a treatments-by-subjects experiment two males are assigned to treatment A and four to B, and four females are assigned to treatment A and two to B. The following experimental results are obtained:

	A		B		
Males	3		5	6	$\bar{X} = 4\ 5/6$
	4		5	6	
Females	8	10	12		$\bar{X} = 10\ 1/2$
	10	11	12		
$\bar{X} =$	7 2/3		7 2/3		

For both males and females the larger scores are associated with the B treatment; yet the A and B means are equal. As the means are equal, $\Sigma(T^2/n)$ has as small a value as could be obtained from any of the data permutations, and so the probability value would be the largest possible, namely, 1.

An overall one-tailed test, on the other hand, using the sum of the B measurements as the test statistic, would give a P-value of 1/225, or about 0.004, because no other data permutation out of the $(6!/2!4!)^2$ data permutations would give such a large sum of B as 46, the obtained value. The sum of the B measurements is an equivalent test statistic, under data permutation, to the arithmetic difference $\bar{X}_B - \bar{X}_A$, and the value of this test statistic is a maximum for the obtained results, although its value is only 0; all other data permutations would have a negative value for the difference between means.

The one-tailed test, then, leads to a reasonable significance value in Example 6.9, but the nondirectional test does not. Of course, if there were more than two treatment levels, some type of nondirectional test statistic related to a difference between means would be used. Yet, as we have seen, such a test might lead to nonsignificant results even when within each type of subject there was an indication of an effect, and even with the same direction of difference between means.

A solution to the problem lies in the development of a more appropriate nondirectional test statistic than $\Sigma(T^2/n)$. The new test statistic is based on a redefinition of SS_B which is mathematically equivalent to the conventional definition when the cell sizes are proportional but which differs when there are disproportional cell sizes: $SS_B = n_A[\bar{X}_A - E(\bar{X}_A)]^2 + n_B[\bar{X}_B - E(\bar{X}_B)]^2 + \cdots$, where $E(\bar{X}_A)$ and $E(\bar{X}_B)$ are "expected means," the mean value of those means over all data permutations. When cell sizes are proportional, the expected means are the same for all treatments, each expected mean having the value of the grand mean of all measurements. Thus, the redefined SS_B is a generalization of the conventional SS_B statistic resulting from substituting expected means for the grand mean in the standard formula, a substitution that has no effect when there are proportional cell sizes but which may have considerable effect when the cell sizes are disproportional.

Let us compute the new test statistic for the data in Example 6.9. $\bar{X}$ for males is 4 5/6 and $\bar{X}$ for females is 10 1/2. $E(\Sigma X_A) = 2(4\ 5/6) + 4(10\ 1/2) = 51\ 2/3$. $E(\Sigma X_B) = 4(4\ 5/6) + 2(10\ 1/2) = 40\ 1/3$. $E(\bar{X}_A)$ is then (51 2/3)/6, or 8 11/18, and $E(\bar{X}_B)$ is (40 1/3)/6, or 6 13/18. We now use these expected means to compute our redefined SS_B test statistic. The obtained test statistic value is computed as $6(7\ 2/3 - 8\ 11/18)^2 + 6(7\ 2/3 - 6\ 13/18)^2$, or about 10.7. $E(\bar{X}_A)$ and $E(\bar{X}_B)$ are, of course, the same for all data permutations, but $\bar{X}_A$ and $\bar{X}_B$ vary. Over all data permutations, however, only the obtained data configuration gives as large a test statistic value as 10.7, and so the significance of the treatment effect by data permutation, using the new nondirectional test statistic, is 1/225, or about 0.004.

6.13 RESTRICTED-ALTERNATIVES RANDOM ASSIGNMENT

With disproportional random assignment, the random assignment has been shown to function in a different manner from what is generally expected. Customarily, we think of random assignment as "equating" treatments with respect to subjects, not necessarily for a particular random assignment but "on the average" over all possible assignments. But when females tend to give larger scores than males under each of two treatments and we assign a small proportion of males and a large proportion of females to treatment A, obviously the treatments have not been equated with respect to subjects, even with random assignment within each sex. Yet such "biased" random assignment in no way affects the validity of data permutation, because the actual type of random assignment with the sample size constraints is taken into consideration in the data permutation. The objective in using a new SS_B test statistic is to increase the *power* of the test; the *validity* of using the same test statistic as with proportional cell sizes is not in question. The possibility of conducting valid statistical tests when the random assignment tends to favor certain treatments more than others opens up new opportunities in experimental design. One of these is what we will call the "restricted-alternatives random assignment" experiment, where the set of treatments taken in a repeated-measures experiment varies over subjects or, for an independent-groups experiment, the set of alternative treatments to which a subject can be assigned varies over subjects.

There are various reasons why subjects might be willing to take some treatments but not others. In an experiment comparing the effects of tobacco, alcohol, and marijuana, some persons may be willing to submit to two of the treatments but not a third and which two they would be willing to take could vary from subject to subject. For some experiments a subject's religious beliefs or moral or ethical attitudes may make certain treatments unacceptable and others acceptable. Similarly, medical reasons justify assigning a certain subject to some treatments but make other assignments inappropriate. If we restrict our experimental designs to the conventional designs in which all subjects must be willing to be assigned to any of the treatments (or, in the case of repeated-measures experiments, designs in which all subjects must be willing to take all of the treatments), we are restricting our potential subject pool. This may not matter for experiments where it is easy to get enough subjects, but it does matter when subjects are difficult to obtain.

Example 6.10

We have an independent-groups experiment with three treatments, tobacco, alcohol, and marijuana, where reaction time is the dependent variable. Three subjects are willing to be assigned to any of the three conditions (T, A, or M), six to be assigned to tobacco or alcohol (T or A), four to tobacco or marijuana (T or M), and two to alcohol or marijuana (A or M). The random assignment was carried out for each of the four groups, with the *sample size* constraints shown in Table 12. What started out as a single-factor independent-groups experiment has become a factorial experiment, a treatments-by-subjects experiment where the subjects are categorized according to their treatment alternatives. Within each level of the subject variable, the subjects are assigned randomly to treatments with the sample size and treatment constraints shown. The disproportionality of cell sizes implies that we should use the new SS_B test statistic to test the difference between treatment means. The alternative treatment assignments are a function of the individual subjects. If older persons (who would be expected to have longer reaction times) were less willing than younger persons to have marijuana as a possible assignment, younger persons would be more likely to be assigned to marijuana, and if the treatments had identical effects one would expect the marijuana reaction times to be lower. So if the column means were equal, it would imply a *differential* treatment effect strong enough to counteract the bias due to subject assignment.

The expected treatment means are computed in the following way, where the numerical subscripts for $\overline{X}$ indicate the row.

TABLE 12   Restricted-alternatives Assignment

	T	A	M
T, A, or M	1	1	1
T or A	3	3	—
T or M	2	—	2
A or M	—	1	1

$E(\Sigma T) = (1)(\bar{X}_1) + (3)(\bar{X}_2) + (2)(\bar{X}_3)$. $E(\bar{T}) = E(\Sigma T)/6$, since 6 is the number of subjects assigned to T. Similarly $E(\Sigma A) = (1)(\bar{X}_1) + (3)(\bar{X}_2) + (1)(\bar{X}_4)$. $E(\bar{A}) = E(\Sigma A)/5$, since five subjects were assigned to A. $E(\Sigma M) = (1)(\bar{X}_1) + (2)(\bar{X}_3) + (1)(\bar{X}_4)$. $E(\bar{M}) = E(\Sigma M)/4$, since four subjects were assigned to M. Then the new SS_B test statistic is computed over all data permutations, of which there are $(3!/1!1!1!) \times (6!/3!3!) \times (4!/2!2!) \times (2!/1!1!) = 1,440$. For comparisons of only two treatments (tobacco and alcohol, for example), only the subjects who took one of those two treatments and who could have been assigned to the other treatment are considered. For instance, for a comparison of tobacco and alcohol, we divide the two measurements for row 1 under T and A between those two columns, and the six measurements for the second row in every way between T and A, and that is all. As there is proportionality for this part of the table it is unnecessary to use the modified SS_B test statistic. Program 6.1 can be used to test the difference between T and A and provides both a one-tailed and a two-tailed test. The program would provide P-values based on a random sample of the $(2!/1!1!) \times (6!/3!3!) = 40$ data permutations that would be the basis of a systematic permutation test. The sampling is done with replacement, so a sample of several thousand of the data permutations can be taken; that is, the data can be permuted several thousand times randomly, even though there are no more than 40 distinct data permutations that would result. With several thousand random permutations Program 6.1 is likely to provide a test that is almost as sensitive to a treatment effect as a systematic test would be. For a one-tailed test the treatment predicted to provide the longer reaction times would be the first column in the two-dimensional layout prepared before entering the data into the computer.

Example 6.10 dealt with the analysis of data from an independent-groups design involving restricted-alternatives random assignment. A repeated-measures design could be analyzed in a similar way as follows.

Example 6.11

Three subjects are willing to be assigned to tobacco, alcohol, and marijuana, six to tobacco and alcohol, four to tobacco and marijuana, and two to alcohol and marijuana. The experimenter uses a repeated-measures design, in which each subject takes all of the treatments which he would be willing to take. Table 13 shows the treatments the individual subjects would take. For each subject there would be random assignment of his

TABLE 13   Repeated-measures Restricted-alternatives Assignment

	T	A	M
T, A, and M	S_1	S_1	S_1
	S_2	S_2	S_2
	S_3	S_3	S_3
T and A	S_4	S_4	–
	S_5	S_5	–
	S_6	S_6	–
	S_7	S_7	–
	S_8	S_8	–
	S_9	S_9	–
T and M	S_{10}	–	S_{10}
	S_{11}	–	S_{11}
	S_{12}	–	S_{12}
	S_{13}	–	S_{13}
A and M	–	S_{14}	S_{14}
	–	S_{15}	S_{15}

treatment times to the treatments he takes. The disproportional cell sizes could, of course, again favor one treatment over another because of the type of subject most likely to take it.

Let us now consider how to compute the expected treatment means. To get $E(\bar{T})$ we take the average measurement of each of the subjects who took treatment T (namely, the first 13 subjects), add those 13 averages, and divide by 13. The expected means for the other two treatments would be computed in a similar fashion.

The test statistic would be the same as for the independent-groups design: $\Sigma\{n[\bar{X} - E(\bar{X})]^2\}$. In this example, we would have $13[\bar{X}_T - E(\bar{X}_T)]^2 + 11[\bar{X}_A - E(\bar{X}_A)]^2 + 9[\bar{X}_M - E(\bar{X}_M)]^2$ as the obtained test statistic value. This test statistic would be computed for all data permutations to determine the proportion that give a test statistic value as large as the obtained value. There would be six possible assignments for each subject taking three treatments, and two possible assignments for each subject taking two treatments; so the number of distinctive assignments for the experiment as a whole would be $6^3 \times 2^{12}$, or 884,736. If only subjects willing to take all three treatments had been used, there would have been only the first three subjects, providing only 216 possible assignments and, therefore, 216 data permutations instead of 884,736, making it a much less powerful experiment. For comparing only two of the treatments the data permuted would be for each subject taking both treatments and just data obtained by those subjects for the two treatments. For example, for a comparison of T and A there would be 2^9 data permutations since, for each of the nine subjects taking both T and A, there are two possible assignments of those treatments to the two treatment times actually used for those treatments by the subject. There is proportionality for this part of the table, so the modified SS_B test statistic is not required. Program 5.1, for repeated-measures ANOVA, can be applied to determine the proportion of the 2^9 data permutations that provide a one-tailed or two-tailed test statistic value as large as the obtained value.

Both independent-groups and repeated-measures experiments have been discussed as factorial experiments where the subjects are stratified according to treatment preferences. There could, however, be restricted-alternatives random assignment where the experimenter decides to restrict the alternatives rather than to leave the choice up to the subject. The analysis would, of course, be the same, because it is the restriction on the alternatives for a subject that affects the analysis, not the reason for the restriction. In a treatments-by-subjects design where an experimenter used children, young adults, and older adults as his levels of the subject variable, he might use levels of treatments for young and older adults that would be danger-ous or otherwise inappropriate for children, and so the assign-ments of children would involve greater restriction than the assignments for the older subjects. Another reason for having different restrictions on the random assignment for some levels than for other levels could be the experimenter's interest in certain cells (treatment combinations) and disinterest in others. For instance, in a two-factor design for examining the effects of intensity of sound and intensity of light on work performance, an experimenter may decide that a high intensity of sound com-bined with a high intensity of light would be so uncomfortable as to distract subjects. Since it is not the distracting effect in which he is interested, he decides to use all three levels of intensity for both sound and light, but to assign no subjects to the ninth cell; his analyses use only eight of the cells in the 3×3 matrix, the ninth cell being empty. There are thus various reasons why there might be restricted-alternatives ran-dom assignment, some of which simply provide disproportionality while others provide empty cells in the factorial design. In either case, the use of the new SS_B test statistic based on expected means permits a valid and powerful analysis.

6.14 DATA ADJUSTMENT FOR DISPROPORTIONAL CELL FREQUENCY DESIGNS

We have considered an adjustment in computation of a test sta-tistic to make a randomization test more sensitive to treatment effects when cell frequencies are disproportional. The superi-ority of the modified SS_B test statistic over $\Sigma(T^2/n)$, the non-directional test statistic in Program 6.1, has been shown for a particular situation (Example 6.9), and the rationale given in Section 6.12 suggests that the superiority holds for designs with disproportional cell frequencies, in general. The rationale, however, concerned analysis of raw data, and it will be shown here that $\Sigma(T^2/n)$ applied to *adjusted data* is just as effective in adapting a randomization test to disproportional cell frequen-cies as is the modified SS_B test statistic applied to raw data.

It will be recalled that the treatments-by-subjects experi-ment in Example 6.9, where sex was the subject variable, gave paradoxical results because of the difference in the proportional distribution of males and females between the two treatments *plus* systematically larger measurements for females under each treatment. If the male and female measurements could be adjusted to make them equal on the average, such an adjustment should serve the same function as adjusting the test statistic.

One way to equate the measurements of males and females is to express all measurements as deviations from the mean for that sex. Thus, since the mean for males was 4 5/6 and the mean for females was 10 1/2, the data in Example 6.9 would be adjusted by subtracting 4 5/6 from each measurement for males and subtracting 10 1/2 from each measurement for females. That adjustment would result in the following array of "residuals":

	A		B	
	$-1\ 5/6$		$+1/6$	$+1\ 1/6$
Males				
	$-5/6$		$+1/6$	$+1\ 1/6$
	$-2\ 1/2$	$-1/2$	$+1\ 1/2$	
Females				
	$-1/2$	$+1/2$	$+1\ 1/2$	
$\bar{X} =$	$-17/18$		$+17/18$	

SS_B for the above results, computed in the usual way where deviations are taken from the grand mean, is $6(-17/18 - 0)^2 + 6(+17/18 - 0)^2$, exactly the same value as the adjusted SS_B in Section 6.12, the value which rounds off to 10.7. This numerical identity is not an accident; the two procedures are mathematically equivalent ways of determining the same value of SS_B. Now, $\Sigma(T^2/n)$ is an equivalent test statistic, under data permutation, to SS_B computed as above, in the conventional manner, and so $\Sigma(T^2/n)$ computed for all permutations of the residuals will give the same P-value as using the adjusted SS_B.

Program 6.1, therefore, will give the same P-value for a nondirectional test as is given by the use of modified SS_B when the data are adjusted in the above manner before entering the data into the computer. Thus, Program 6.1 can serve as a sensitive test of treatment effects for either proportional or disproportional cell frequency designs. Although the discussion of the need to adjust for disproportionality of cell frequencies has been in terms of treatments-by-subjects designs, the same considerations apply to other randomized block designs and to completely randomized designs. Data adjustment of the above kind can be helpful for any disproportional cell frequency design. Program 6.1, however, cannot be applied to the special type of design in which, as in restricted-alternatives designs, there are empty cells, because such designs require a different data permuting procedure.

The above procedure of computing the residuals once and then permuting the residuals provides the same distribution of test statistics as the more cumbersome procedure of permuting the raw data and computing the residuals for each permutation of the data. The reason is that the residual associated with each measurement is not affected by permuting within rows.

6.15 DESIGNS WITH FACTOR-SPECIFIC DEPENDENT VARIABLES

A factorial design can be arranged so that the dependent variable is different for different factors. As in other factorial designs, the experimental units would be assigned randomly to levels of two or more factors but instead of taking measures on a single dependent variable the experimenter takes measures on several dependent variables, one for each of the factors.

Example 6.12

An experimenter is interested in the effect of a lesion in location B in a rat's brain on the rat's brightness discrimination, and he is interested also in the effect of a lesion in location L in the rat's brain on the rat's loudness discrimination. To make the optimum use of data from a small number of rats, and to minimize time for performing the operations and allowing the rats to recover, the experimenter decides to use the same rats for both experiments and to make any required lesions during the same operation. He assigns three rats at random to each of the four treatment conditions represented by the four cells in the following matrix:

		Lesion in location B?	
		Yes	No
Lesion in	Yes	n = 3	n = 3
location L?	No	n = 3	n = 3

Three rats have both lesions, three have no lesions, three have a lesion only in B, and three have a lesion only in L.

After all of the lesioned rats have recovered from the operation, all of the rats are tested on brightness and loudness discrimination tasks which they have learned previously. Thus each rat gets a measurement for two dependent variables: brightness discrimination and loudness discrimination.

The experimenter evaluates the results of the lesion in location B with only the brightness discrimination measurement and the results of the lesion in location L with only the loudness discrimination measurement. For the effect of lesion B the H_0 would be: whether or not a rat had a lesion in location B had no effect on its brightness discrimination measurement. There would be 400 data permutations used to test this H_0, consisting of those associated with each of the $6!/3!3! = 20$ divisions of brightness discrimination data in the top row between the two cells in conjunction with each of the 20 divisions of the brightness discrimination data in the bottom row between the two cells. For this test, the loudness discrimination data would be ignored. The test statistic could be $\Sigma(T^2/n)$ for a nondirectional test, or it could be the sum of the brightness discrimination measurements in the column expected to provide the larger measurements, for a one-tailed test. A test for the effect of lesion L on loudness discrimination would be conducted in an analogous manner, dividing the loudness discrimination measurements between cells within columns to generate the set of data permutations. Program 6.1 can be used to carry out these tests.

6.16   DICHOTOMOUS AND RANKED DATA

Throughout this chapter the issues have concerned all types of quantitative data. Consequently, what has been said so far is applicable also to cases where the responses fall into one of two categories, because the dependent variable "measurement" can then be expressed as 0 or 1 to apply the randomization test procedure. If the data are in the form of ranks, there are two options. One is to rank within rows separately for testing for column effects and within columns for testing for row effects. The other option is to rank over all cells and to permute those ranks within columns or rows in testing for treatment effects. The latter procedure should be more sensitive to treatment effects as it provides for finer discriminations in much the same way as unranked data permits more subtle discriminations than ranks. Consider the following matrices of ranks:

	A_1	A_2		A_1	A_2
B_1	3, 4	5, 8	B_1	1, 2	3, 4
B_2	1, 2	6, 7	B_2	1, 2	3, 4

Consider a test of the effect of factor A. Notice that in the matrix on the left, showing ranks over all cells, switching ranks 4 and 5 has less effect on a one- or two-tailed test statistic than switching ranks 2 and 6; yet, with reranking within rows, as shown in the right matrix, the corresponding switches, namely of ranks 2 and 3 within B_1 or within B_2 levels, would have the same effect. But clearly, as the ranks 4 and 5 are between 2 and 6, they represent effects which are not as different and thus reversals might more readily occur by chance than for ranks 2 and 6.

6.17 PROBABILITY COMBINING

Probability combining can be employed when different tests are statistically independent. The required statistical independence is: when H_0 is true for each of the tests for which the P-values are combined, the probability of getting a P-value as small as p for one of the tests must be no greater than p, no matter what P-value is associated with any of the other tests. The method of probability combining to be described in Section 6.18 uses the sum of the P-values to determine the overall significance of the results of the statistical tests. The general H_0 tested by probability combining is: there is no treatment effect for any of the tests that provided probabilities to be combined.

One type of factorial design where probabilities can be combined is that where the probabilities come from separate experiments. The separate experiments could be a series of pilot studies and the main experiment. Pilot studies are used to determine how to conduct the main experiment. Not only the general conduct of the experiment but also the independent and dependent variables may be based on preliminary pilot work. Observations in a pilot experiment may suggest changes likely to make an experiment more sensitive even when the pilot experiment is not sensitive enough to give significant results. Probability combining permits the experimenter to use pilot studies not only for designing a final experiment but also to provide P-values that can be combined with the P-value from the main experiment.

Example 6.13

A pilot study was run preliminary to a planned study of the effect of intensity of lighting on the amount of food rats eat. A statistical test suggested an effect of light intensity on the amount of food eaten. A record of the variety of foods eaten was not kept, but incidental observation suggested a strong effect of light intensity on the number of different foods eaten. Thus, in designing the main experiment, two dependent variables are used: the total amount of food eaten and the number of different foods eaten. Different rats from those used in the pilot study are randomly assigned to the treatments, and P-values are obtained from the main experiment for both dependent variables. The two P-values for the two dependent variables in the main experiment cannot be combined because they are not independent: in the absence of an effect on either dependent variable there could be a correlation between the two P-values because of a high correlation between the dependent variables. The P-values from the pilot and main experiments for tests using amount of food eaten as the dependent variable could, however, be combined as they would be statistically independent. The experimenter then would have a relatively sensitive determination of whether light intensity affected the amount of food eaten and, if the main experiment showed the strong effect on varying the foods eaten that was suggested by the pilot study, the one P-value for the variety of food eaten in the main experiment should reveal that treatment effect. In other words, the experimenter could test the effect on his powerful dependent variable in the main experiment and could combine the P-values for the two tests involving the weaker dependent variable.

The P-values combined over pilot studies need not concern identical dependent variables. An experimenter may choose to modify the dependent variable on the basis of pilot work.

Example 6.14

An experimenter wants to determine whether injection of a certain drug makes monkeys irritable. There are only a small number of monkeys available for the experiment, so the experimenter has to have an efficient design. He decides to use one-third of the monkeys in a pilot study to determine the most sensitive dependent variable and the other two-thirds in the second experiment, employing that dependent variable. He expects irritability to be manifested mainly in destructive behavior, and so in the pilot study he records the number of instances of destructive behavior as the dependent variable. It appears that the animals given the drug were more inclined to be destructive but the evidence is not strong. However, it seems that the drug made the monkeys more likely to be hostile toward other monkeys. Thus, for a second experiment, using the remaining monkeys, the experimenter uses a measure of hostility as the dependent variable. The P-values from the two experiments then are combined. The two dependent variables are not identical but both could be considered measures of irritability. The sum of the two P-values may be sufficiently small to permit the experimenter to reject the null hypothesis of no effect of the drug on irritability.

A different application of probability combining over separate experiments is that associated with a planned series of experiments using the same subjects in different experiments. Use of the same subjects is not always appropriate but, with independent random assignment, valid and independent statistical tests can be performed on data from the same subjects in different experiments. Randomization test P-values based on the same subjects in different experiments are statistically independent if there is independent random assignment to treatments in the different experiments. Participation in an earlier experiment may very well affect performance in a later experiment, in which case the *responses* in the second experiment would not be independent of the responses in the first experiment; but it is the independence of the P-values when H_0 is true that is of importance. If the first experiment made the subjects responsive to the difference between treatments in the second experiment, it could affect the P-values in the second experiment. This effect would not invalidate probability combining however, because in such a case H_0 for the second experiment would be *false* and so the probability of rejecting a *true* H_0 at any level α still would be no greater than α. Separate random assignment ensures the independence of the *P-values* if the generic null hypothesis of no treatment effect in either (any) of the experiments is true.

Let us reconsider the experiment in Example 6.13, in which the experimenter wanted to determine the effect of intensity of light on the amount of food eaten. If the experimenter had only five subjects, the smallest P-value possible for a single experiment would be 0.10, with two subjects taking one treatment and the other three taking the other treatment, since there are only 10 possible assignments. However, the experimenter could conduct the same experiment three times in succession with the same five subjects, randomly assigning the subjects on each occasion. The P-values would be independent and pooling over the three experiments might conceivably permit the experimenter to reject H_0 at the 0.01 level.

6.18 ADDITIVE METHOD OF COMBINING PROBABILITY VALUES

There are two main procedures for combining probability values, which can be designated as the *multiplicative* and the *additive* methods. The multiplicative method determines the significance of the product of the probability values to be combined, whereas the additive method determines the significance of the sum of the probability values. Both are valid methods for employment with independent P-values, but only the additive method will be described here.

If we take the sum of n independent P-values, then, given that H_0 associated with each P-value is true, the probability of getting a sum as small as S is:

$$\frac{C(n,0)(S - 0)^n - C(n,1)(S - 1)^n + C(n,2)(S - 2)^n - \cdots}{n!}$$

$$(6.1)$$

The minus and plus signs between terms in the numerator alternate, and additional terms are used as long as the number subtracted from S is less than S (Edgington, 1972). The symbol $C(n, r)$ is the same as $n!/r!(n - r)!$ and refers to the number of combinations of n things taken r at a time.

Example 6.15

Combining as many as seven different probabilities will not be required very often, but such an example will be helpful in showing how to use formula (6.1). Suppose that seven experiments provided P-values of 0.20, 0.35, 0.40, 0.40, 0.35, 0.40, and 0.40. The sum is 2.50. The probability of getting a sum as small as 2.50 from seven independent P-values when each of the seven null hypotheses is true is:

$$\frac{C(7,0)(2.50 - 0)^7 - C(7,1)(2.50 - 1)^7 + C(7,2)(2.50 - 2)^7}{7!}$$

$$= 0.097$$

On the other hand, if the seven values had been somewhat smaller, providing a sum of 1.40, the overall probability or significance value would have been:

$$\frac{C(7,0)(1.40 - 0)^7 - C(7,1)(1.40 - 1)^7}{7!} = 0.002$$

Since no single P-value can exceed 1, the maximum possible sum of n P-values is n, and so there never needs to be more than n terms computed. Thus, for three P-values to be summed, no more than three terms would be required, and for four P-values no more than four terms would be required.

6.19   COMBINING TWO-TAILED PROBABILITIES

We will distinguish now between combining one- and two-tailed P-values. When all of the P-values to be pooled are one-tailed P-values, the matter is simple: use the P-values for the predicted direction of difference; if the obtained results differ in the predicted direction, the P-value may be fairly small, but if the obtained results differ in the opposite direction the P-value will be large.

When P-values are two-tailed, there are two alternative procedures that can be followed, depending on the predicted effect. First, consider the case where the experimenter has no basis for predicting the direction of difference for any of the experiments and does not even have reason to expect that the unknown direction of difference will be the same over all experiments. In such a case he should simply combine the two-tailed P-values.

In the second situation, suppose that the experimenter has no basis for predicting the direction of difference in any of the experiments, but that he does expect the direction of difference to be the same for all experiments. In this case, the

experimenter should construct two sets of P-values, one set
based on one-tailed P-values for one direction of difference
and the other set based on one-tailed P-values for the opposite
direction of difference. There are two sums, one from one set
and one from the other set. Significance of the combined results is based on the smaller sum, but the computed significance
value for the smaller sum is doubled.

Example 6.16

We have three experiments which, although not identical, all
involve two treatments that are analogous from one experiment
to the next. One treatment for each experiment may be called
"pleasant" and the other "unpleasant." The experimenter expects
the treatment giving the larger mean to be the same one in every
experiment, but does not know whether it would be the
"pleasant" or the "unpleasant" treatment. The one-tailed
P-values, based on a prediction that the "pleasant" treatment
measurements would be larger, are: 0.10, 0.08, and 0.16.
The sum of this set is 0.34. The sum of the set of one-tailed
probabilities based on the opposite prediction is, of course, considerably larger. The value $(0.34)^3/6$, which is about 0.0066,
is doubled to give 0.013 as the significance of the combined
results.

Suppose the two-tailed probabilities were twice the one-
tailed probabilities for the correctly predicted direction. (They
need not be, for randomization tests, in cases of unequal sample
sizes.) If they had been combined without taking into consideration the expectation that the direction of difference would be
the same for all three experiments, we would have combined the
values 0.20, 0.16, and 0.32 to get a total of 0.68 and would
have obtained a significance value of $(0.68)^3/6$, or about 0.052.
The difference between 0.052 and 0.013 shows the importance
of taking into account a prediction of a consistent direction of
difference in combining two-tailed probabilities, even when the
direction cannot be predicted

This method of combining two-tailed probabilities was recommended by David (1934) and is based on the same principle as
Fisher's modified least-significant difference procedure, described
in regard to multiple comparisons in Sections 4.11 and 5.10.
The principle is this: when H_0 is true for each of k different
tests on the same or different sets of data, the probability of
significance at the α level is no greater than $k\alpha$. David's method
of combining two-tailed probabilities carries out two "tests" on

the results from a set of experiments, and the probability of
one of the tests giving significance at the α level is no greater
than 2α.

Further discussion of the combining of probabilities can be
found in reports by Baker (1952), Bancroft (1950), Gordon et
al. (1952), Guilford and Fruchter (1973, pp. 210–212), Jones
and Fiske (1953), Lancaster (1949), Wallis (1942), and Winer
(1971, pp. 49–50).

6.20   REPORTING RESULTS OF STATISTICAL TESTS

In Chapters 4 and 5, the statistical tests were conventional ones
with significance determined by the use of randomization tests.
The research report of the statistical analysis thus could be
standard except for indicating that the significance of t or F
was determined by means of a randomization test and explaining
why t or F tables were not used to determine significance.   The
present chapter, on the other hand, concerns a number of tests
and test statistics without familiar parametric counterparts,
making a research report of the results more difficult.

Many of the experimental designs described in this chapter
are those for which factorial ANOVA is commonly used.   In those
cases it is helpful to the reader to use familiar terms in referring
to those designs, such as *factor, factor levels, randomized
blocks*, and *main effects*.   The use of such terms does not imply
that factorial ANOVA was performed; a factorial *design* does not
require the computation of F or an equivalent test statistic for
the statistical analysis.   Program 6.1 is a computer program for
a randomization test for factorial designs but not for factorial
ANOVA.

An experimenter should describe the random assignment
procedure employed in his factorial experiment, the null hypothesis, and the test statistic involved in the randomization test.
(If Program 6.1 is used, it would be better to describe the test
as employing SS_B as a test statistic, since the test statistic
actually computed, $\Sigma(T^2/n)$, would be unfamiliar but equivalent
to SS_B for the randomization test.)   The number of random data
permutations also should be specified.

The discussion in this chapter of validity and the considerations involved in combining one- and two-tailed probability values
apply also to the multiplicative method given in some books

(e.g., Winer, 1971, pp. 49–50), wherein use is made of chi-square and natural logarithm tables to determine significance. The computation of the overall significance value need not be given to the reader, but the interpretation should be provided. For instance, if the overall combined probability value from several experiments is significant at the 0.01 level, it could be explained that no more than one percent of the time would such a small sum (or in the case of the multiplicative method, such a small product) of P-values be obtained when H_0 is true for each of the experiments. It might be appropriate also to state the alternative hypothesis, the hypothesis that there was a treatment effect for some (one or more) of the experiments, so that the reader does not assume that rejection of the H_0 implies a treatment effect in *all* of the experiments. (If the experiments are very similar, such a conclusion might *logically* follow from rejection of H_0, but there is no *statistical* justification for generalizing to all experiments.) When separate experiments involving the same subjects provide P-values to be combined, reference to the independence of the random assignment of the subjects from one experiment to the next should be given to justify the assumption of independence of the P-values.

Whether analysis of data from designs with disproportional cell frequencies is carried out by the use of the new SS_B test statistic or by the application of Program 6.1 to residuals, the P-value will be the same. Thus, the procedure which is easier to explain should be presented in the research report as the procedure employed.

When restricted-alternatives random assignment has been used and the significance determined by the randomization test procedure, it is desirable to describe the random assignment procedure and to explain that the data were permuted in a corresponding way to ensure validity of the randomization test. Since a randomization test that permutes in this way will give the same P-value whether it permutes raw data and computes the new SS_B test statistic each time or permutes residuals and computes $\Sigma(T^2/n)$ or the conventional SS_B as the test statistic for the residuals, the results may be described in terms of whichever approach is easier to describe. If the use of residuals is the way the procedure for determining significance is described, the discussion should be in terms of SS_B, even though the simpler but equivalent test statistic $\Sigma(T^2/n)$ was computed instead of SS_B for each data permutation.

REFERENCES

Baker, P. C. (1952). Combining tests of significance in cross-validation. *Educ. Psychol. Meas. 12*, 300–306.

Bancroft, T. A. (1950). Probability values for the common tests of hypotheses. *J. Am. Stat. Assoc. 45*, 211–217.

David, F. (1934). On the $P_{\lambda n}$ test for randomness; remarks, further illustrations and table for $P_{\lambda n}$. *Biometrika, 26*, 1–11.

Edgington, E. S. (1969). *Statistical Inference; the Distribution-free Approach*. McGraw-Hill, New York.

Edgington, E. S. (1972). An additive method for combining probability values from independent experiments. *J. Psychol. 80*, 351–363.

Gordon, M. H., Loveland, E. H., and Cureton, E. E. (1952). An extended table of chi-square for two degrees of freedom, for use in combining probabilities from independent samples. *Psychometrika 17*, 311–316.

Guilford, J. P., and Fruchter, B. (1973). *Fundamental Statistics in Psychology and Education* (5th ed.). McGraw-Hill, New York.

Jones, L. V., and Fiske, D. W. (1953). Models for testing the significance of combined results. *Psychol. Bull. 50*, 375–382.

Lancaster, H. O. (1949). The combination of probabilities arising from data in discrete distributions. *Biometrika 36*, 370–382.

Wallis, W. A. (1942). Compounding probabilities from independent significance tests. *Econometrica 10*, 229–248.

Winer, B. J. (1971). *Statistical Principles in Experimental Design* (2d ed.). McGraw-Hill, New York.

7
Multivariate Designs

Factorial designs are designs with two or more *independent* variables. Multivariate designs, on the other hand, are designs with two or more *dependent* variables. Treatment effects in experiments are likely to be manifested in a number of ways, and so it frequently is useful to employ more than one dependent variable. For example, an experimenter may expect that under one condition a rat would show greater motivation to solve a task than under another and that this greater motivation would be reflected both in the speed and the strength of response. If he measures only one of the two aspects of the response in testing for treatment effect, his experiment is likely to be less sensitive than if both measures were incorporated into the experiment, because one measure may reflect an effect when the other does not. By allowing for the use of two or more dependent variables in a single test, multivariate designs are thus more sensitive to treatment effects than are univariate (single-dependent-variable) designs.

7.1 NEED FOR MULTIVARIATE RANDOMIZATION TESTS

Multivariate analysis of variance (MANOVA) is a common statistical procedure for analyzing data from multivariate experimental designs. There are several conventional test statistics for multivariate ANOVA, and these alternative test statistics sometimes provide quite different significance values. Olson (1976) studied the robustness of MANOVA and found that violations of the parametric assumptions have a considerable effect on the probability of incorrectly rejecting the null hypothesis for several of the commonly used test statistics. (This finding supports a widely held view that parametric assumptions are much more important for multivariate than for univariate, or one-variable, ANOVA.) Olson constructed three populations of numerical values with identical means on three dependent variables. For one population the standard deviation for each dependent variable was three times the standard deviation for that variable in the other two populations. A random sample of measurements for five subjects was taken from each population, and each of four MANOVA test statistics was computed with the significance being determined by significance tables. The sampling and the significance determination were done repeatedly, and the proportion of the test statistics of each kind significant at the 0.05 level was computed. This proportion ranged from 0.09 to 0.17 for the four test statistics. (The proportion should, of course, be no greater than 0.05 if the method of determining significance was valid.) A similar study was performed with six dependent variables, where one population had dependent variable standard deviation six times as large as for the other two populations. The proportion of the time that a test statistic was judged to be significant at the 0.05 level ranged from 0.09 to 0.62 for the four test statistics. A test statistic known as the Pillai-Bartlett Trace V, along with its conventional procedure for determining significance, was the most robust, providing the proportion of 0.09 for both studies. Olson's results, considered in conjunction with other theoretical and numerical demonstrations of the sensitivity of MANOVA procedures to violations of assumptions, show the need for a valid procedure of determining significance.

The implications of Olson's study for situations where no populations have been randomly sampled is unclear, but results of his and similar studies on the sensitivity of MANOVA procedures to violations of parametric assumptions tend to make people more critical of the validity of MANOVA. A randomization test determination of significance for MANOVA thus may increase the acceptability of results at times.

7.2   RANDOMIZATION TESTS FOR CONVENTIONAL MANOVA

Apparently the most common approach in using a multivariate t test or MANOVA is equivalent to combining measurements on different dependent variables into single numbers to which uni-variate tests are applied. Harris (1975, p. xi) stated: "Each of the commonly employed techniques in multivariate statistics is a straightforward generalization of some univariate statistical tool, with the single variable being replaced by a linear combination of several original variables." The set or "vector" of measurements for a subject is reduced to a single numerical value by determining the "optimum" weight by which to multiply each value and then adding the weighted values. Hotelling's T^2 test, for example, is, in effect, a procedure of combining dependent variable measurements into a single measurement for each experimental unit before carrying out a t test on the composite measurements. (Harris, 1975, pp. 13–14.)

Consider a randomization test for MANOVA for a single-factor completely randomized (independent-groups) design in which N subjects are assigned randomly to three treatments. Two dependent variable measurements are obtained for each subject. The obtained test statistic value then is computed for the obtained results by using a procedure of deriving a single composite measurement from the vector of measurements for each subject and applying one-way ANOVA to compute F. Then in permuting the data, the vectors of original measures are permuted and, for each data permutation, the vectors are reduced to composite measurements, which are used to compute F. Computing the linear combination of values on the different variables for each data permutation is more time consuming than computing the composite measurements only once and permuting the composite measurements each time, but randomization tests require using the slower procedure. If permuting composite measurements was acceptable, the computation for the obtained results could be performed to provide composite measurements to which an ANOVA program from Chapter 4 could be applied. Unfortunately, permuting the composite measurements is not acceptable. The composite measurements are derived by a combining procedure designed to maximize the separation of groups for the obtained results, and thus the composite measurement values vary as the vectors of the original measurements are permuted among groups. The permuting of composite measurements here is unacceptable for the same reason that permuting of residuals for analysis of covariance and the permuting of residuals based on column and row means for a test of interaction were unacceptable: the adjustment of the data depends on the particular data permutation that represents the obtained results.

The same considerations apply to other designs where a randomization test counterpart of a conventional multivariate test is desired. A computer program would require repeated computations of composite measurements and permuting of vectors from which the composite measurements are computed; thus, such a program might take considerable time to run.

7.3 EFFECT OF UNITS OF MEASUREMENT

As pointed out above, when conventional MANOVA procedures are used, in effect, measurements from several dependent variables are combined into a single overall measure of treatment effect and then the single composite measurements are analyzed by ordinary univariate statistical procedures. Alternatives to conventional procedures can be employed in deriving composite measurements, but before considering them, let us examine the units of measurement problem, a matter of importance for any type of composite measurement.

The combining procedure is more likely to increase the sensitivity of the test if we use a procedure that is not affected by units of measurement. Familiar parametric procedures and all of the statistical tests considered so far in this book are such that the significance value is independent of the unit of measurement. The independence of the significance value and the unit of measurement also is very desirable in using composite measurements, because we do not want the significance to depend on such an arbitrary consideration as whether measurements are expressed in inches or millimeters for one of the dependent variables.

Example 7.1

We have the following measurements of snakes under treatments A and B, where three snakes take each treatment and the length and weight of snakes are measured.

	A				B		
	Inches	Ounces	Total		Inches	Ounces	Total
a	12	16	28	d	36	48	84
b	24	62	86	e	60	80	140

	A				B		
	Inches	Ounces	Total		Inches	Ounces	Total
c	48	30	78	f	72	98	170
		T_A = 192				T_B = 394	

For the test, the six "totals" or composite measurements would be allocated in all 6!/3!3! = 20 ways between the two treatments, and T_B would be computed for each data permutation for a one-tailed test where B was expected to provide the larger measurements. Of the 20 data permutations, two provide a T_B as large as 394, the obtained value, so that the one-tailed P-value would be 0.10. Now let us see what would have happened if we had expressed the length in feet and the weight in pounds:

	A				B		
	Feet	Pounds	Total		Feet	Pounds	Total
a	1	1	2	d	3	3	6
b	2	3.875	5.875	e	5	5	10
c	4	1.875	5.875	f	6	6.125	12.125
		T_A = 13.750				T_B = 28.125	

In this case, there is only one permutation giving as large a value of T_B as 28.125, the obtained value, and so the significance value would be 0.05. Thus with the same data, using different units of length and weight, different significance values can be obtained if the numerical values are combined to give the composite measurement.

The problem of controlling for units of measurement in conventional MANOVA linear combinations of measurements is solved by the procedure for determining the weights to assign the measurements from different dependent variables. If we have two dependent variables X and Y, we determine weights w_1 and w_2 such that $w_1X + w_2Y$ gives composite measurements that are "best" in the sense of, say, providing the largest F for the obtained results when F is applied to the composite measurements.

Suppose it turned out that equal weighting was determined to
be best in the snake example for measurements when length is
expressed in feet and weight in pounds. Then if length was
transformed to inches and weight to ounces, the w_1 and w_2
weighting of the inches and pounds measurements would be in
the ratio of 12:16 instead of 1:1, but the MANOVA test statistic
for any data permutation would be the same.

7.4   MULTIVARIATE TESTS BASED ON COMPOSITE z SCORES

The unit of measurement problem can be handled in other ways
of combining measurements from different dependent variables.
The composite z score approach which follows also controls for
units of measurement, and it does have some advantages over
conventional MANOVA test procedures from the standpoint of
performing randomization tests.

As mentioned earlier, the weights have to be computed
separately for each data permutation in performing a randomization test to determine significance for a conventional MANOVA
procedure. For the composite z score approach, on the other
hand, the composite measurements, although dependent on the
joint distribution of measurements for the dependent variables,
are independent of the way the measurements are divided among
treatments. Consequently, the composite measurements can be
computed for the obtained results and then be permuted instead
of being computed for each data permutation.

The composite z score procedure involves transforming the
measurements into z scores, and the z scores for a subject are
added to get a composite z score. The formula

$$z = \frac{X - \bar{X}}{[\Sigma(X - \bar{X})^2/n]^{1/2}} \qquad (7.1)$$

is used to determine the z score for the measurement of a subject
on one of the dependent variables. A z score is not affected by
the unit of measurement, and so a composite z score, and consequently the significance of the sum of the composite z scores,
is independent of the unit of measurement. The measurements
in Example 7.1 will now be transformed to composite z scores.

Example 7.2

The z score for the length of subject b, based on measurement in *feet*, is $(2 - 3.5)/(17.5/6)^{1/2} = -0.88$, and the z score for the length of subject b, based on measurement in *inches*, is $(24 - 42)/(2,520/6)^{1/2}$, which also equals -0.88. The following table shows z-score values for the lengths and weights, and the composite z scores, no matter what the units of measurement may be:

	A		
	Length	Weight	Composite z scores
a	−1.46	−1.41	−2.87
b	−0.88	+0.23	−0.65
c	+0.29	−0.92	−0.63
		$T_A = -4.15$	

	B		
	Length	Weight	Composite z scores
d	−0.29	−0.27	−0.56
e	+0.88	+0.87	+1.75
f	+1.46	+1.51	+2.97
		$T_B = +4.16$	

For determining significance by data permutation, the six composite z scores constitute the "data" that are divided between the two treatments. For each of the 20 possible data permutations, the one-tailed test statistic T_B is computed. It can be seen that no other data permutation provides as large a value of T_B as +4.16, the obtained value, and so the one-tailed P-value is 1/20, or 0.05. (Program 4.4 or 4.5, for the

independent t test, can be used with this design for determining significance.) The P-value happens to be the same as for
the use of the raw measurements expressed in terms of feet and
pounds but is smaller than for the raw measurements expressed
in terms of inches and ounces. Composite z scores are used
to prevent an arbitrary choice of units of measurement from
influencing the significance value; therefore they should tend
to provide more significant results when there is a treatment
effect than if unweighted raw measurements for particular units
of measurement were pooled.

By using composite z scores to test the effect of the two
treatments on length and weight, the variables length and weight
are implicitly regarded as equally important; they make equal
contributions to the composite measurement. If, instead of using
z scores, we expressed each subject's length in inches and its
weight in pounds, and if we used the total of the two numbers
as a subject's composite measurement, almost all of the variation
in the composite measurements would be due to the variation in
length; the weight variable would have little effect. Similarly,
if a composite measurement that was the sum of feet and ounces
were used, variation in the composite measurement would be
primarily a function of variation in the weight variable, and so
length would have little influence on the P-value. The transformation of the lengths and weights to z scores before combining
to obtain a composite measurement serves the function of equating
the variability of the two dependent variables so that they have
equal influence on the significance of the results.

This procedure of generating composite scores by adding z
scores is appropriate whenever the relationship between the
effects on the different dependent variables is predictable. The
above example assumes that the experimenter expected increases
in length generally to go with increases in weight; otherwise, it
would be inappropriate to add the z scores for the two dimensions.
Suppose that the experimenter had three dependent variables, X,
Y, and Z, and that he expected X and Y to change in the same
direction but expected Z to change in the opposite direction.
Z could be the *time* it takes a rat to run a maze in a situation
where the *speed* was expected to vary directly with dependent
variables X and Y. In such a case, the signs associated with
the z scores for Z would be reversed to make them reflect speed
rather than running time. Then the z scores for a subject would
be added to give his composite z score.

Two or more treatments with any number of dependent variables can be handled in this manner. When only two treatments are involved, a one- or two-tailed test can be employed. To determine significance by data permutation for a one-tailed test, the test statistic is the sum of the composite z scores for the treatment predicted to have the larger measurements. (In Example 7.2, treatment B was predicted to have the larger measurements, and so T_B, the total of the composite z scores for treatment B, was used for the test.) For a two-tailed test, the test statistic would be the same as for F for one-way ANOVA: $\Sigma(T^2/n)$. (For Example 7.2 the obtained two-tailed test statistic would be $T_A^2/n_A + T_B^2/n_B$, which would be $(-4.15)^2/3 +$ $(+4.16)^2/3$ for the composite z scores.)

Example 7.3

Independent groups were considered in Example 7.2, but the composite z score approach also can be taken with repeated measures designs when there are several dependent variables. For the independent-groups design, the z score for a subject on a particular dependent variable is based on the mean and standard deviation of the measurements for that variable over all subjects and treatments. The same is true of the z score of a subject on a particular dependent variable for a treatment in a repeated-measures experiment. Subjects S_1, S_2, and S_3 each take three treatments, A, B, and C, the treatment time for each treatment being randomly determined independently for the three subjects. Measurements are made on three dependent variables, X, Y, and Z. The results are shown in Table 1. The measurements for each dependent variable are transformed into z scores. The z scores for all nine X measurements are based on the mean and standard deviation of those nine measurements and, similarly, the z scores of the other two dependent variables are based on the mean and standard deviation of the nine measurements for the dependent variable. For example, the z score for S_1 on dependent variable Z is $(16 - 17.44)/3.69 = -0.39$, where 17.44 is the mean and 3.69 is the standard deviation of the nine Z measurements. A z score is determined for each dependent variable for all subjects, and then the z scores for each subject for a single treatment are added to get a composite z score. For instance, subject S_1 would have three composite z scores: one each for treatments A, B, and C. The composite z score for S_1 for treatment A is the sum of -0.39

TABLE 1   Data from a Repeated-measures
Multivariate Experiment

	Treatment A			Treatment B			Treatment C		
	X	Y	Z	X	Y	Z	X	Y	Z
S_1	5	26	16	8	25	14	12	30	21
S_2	8	23	13	9	24	19	10	25	23
S_3	4	20	12	7	23	18	10	26	21

and the z scores for the measurements on the other two dependent variables. The test statistic is ΣT^2, where T is the total of three composite z scores for a treatment. This test statistic is equivalent to F for repeated-measures ANOVA. There are $(k!)^n = (3!)^3 = 216$ possible assignments of treatment times to treatments for the subjects. The nine composite z scores constitute the data which are permuted. The 216 data permutations consist of every allocation of the three composite z scores for S_1 to the three treatments, in conjunction with every allocation of the three composite z scores for each of the other two subjects. The proportion of test statistic values that are as large as the obtained value is the probability associated with the experimental results. Program 5.1 or 5.2 could be used to determine significance.

It would be possible to compute z scores for a subject on the basis of the mean and standard deviation of the three measurements for the subject on a particular dependent variable. To illustrate, we could compute the z score for the measurement of 16 for S_1 on the basis of the mean and standard deviation of the measurements 16, 14, and 21, for S_1. Such a procedure, however, would be inconsistent with testing significance with univariate repeated-measures designs. As mentioned in Chapter 5, the test statistics t and F for repeated-measures designs are such that subjects with more variability of measurements over treatments have more effect on variation in the test statistic over data permutations than less variable subjects. To compute composite z scores on the basis of each subject's mean and standard deviation for measurements on a particular dependent variable would tend to give subjects the same weight in the determination of the test statistic value.

When the dependent variable values are not based on ordinary measurement but on ranking, there is no need for transformation to z scores, because the rank values for the different dependent variables have equal variability and make composite measurements based on a sum of ranks appropriate, equalizing the influence of all of the dependent variables on the composite measurements. Significance could be determined with Program 4.1 or 4.3 for independent-groups designs or with Program 5.1 or 5.2 for repeated-measures designs.

The composite z score approach requires that a person predict the way the dependent variables covary over the treatments, and to adjust the sign of z scores for a dependent variable accordingly. Useful though this approach may be when there is a basis for judging how dependent variables will covary, a different technique is required when there is no basis for such a judgement. Such a technique is the combining of values of t or F over the different dependent variables.

7.5   COMBINING t OR F VALUES OVER DEPENDENT VARIABLES

Instead of reducing multivariate data to a univariate form by deriving a single composite measure to replace vectors of measurements, one can compute univariate t or F for each dependent variable and use the sum of t's or F's for all dependent variables as a compound test statistic. Thus, to carry out a randomization test on data from a repeated-measures experiment with three dependent variables, F for each of the three variables is computed separately and their sum is used as the test statistic. Although F's are computed separately for the dependent variables, the vectors, not the individual measurements, must be permuted for the test, and a sum of F's must be computed for each data permutation because H_0 concerns the vector of measurements. As the unit of measurement has no effect on t or F, there is no need to transform the data to z scores to control for the unit of measurement.

One feature of this procedure that is unappealing is that with only two treatments the significance given by using the sum of F's as a test statistic may not be the same as that given by using the sum of absolute t values. For instance, with two dependent variables, with absolute t's of 1 and 4, the sum is 5, the same as for a data permutation with t's of 2 and 3, but the sum of the F's (i.e., the square of the t values) for those data permutations would be 17 and 13. One solution to this problem

would be to use the sum of logarithms of F or t as the test statistic. Then the test statistic for F always would be twice that
for absolute t for all data permutations, ensuring the same significance value for the two test statistics. This modification
would have the effect of changing the procedure to one based
on the product, rather than the sum, of test statistics.

7.6   A GEOMETRICAL CONCEPTION OF MANOVA

MANOVA has been approached in two ways in the preceding
sections: as a procedure of reducing vectors of measurements
to single numbers to which univariate ANOVA can be applied,
and as a procedure of applying univariate ANOVA repeatedly to
each dependent variable and summing the obtained F's. A third
approach, based on a spatial or geometrical conception of
MANOVA, follows.

   Various distances in a geometrical conception of MANOVA
are squared in the computation of sums of squares. Squared
distances from *centroids* are computed. A centroid is a multidimensional mean, the vector of the means of all dependent variables. In a scattergram of measures of subjects on variables X
and Y, for example, the point of intersection of a line perpendicular to the X axis at $\overline{X}$ and a line perpendicular to the Y axis
at $\overline{Y}$ is the centroid of the values in the bivariate array.

   The distances within a p-dimensional space for vectors of
measurements from p dependent variables will be treated as
Euclidean distances, where the squared distance between any
two points (vectors) is the sum of the squared distance between
them with respect to each of the dependent variables. For
instance, the squared "straight-line" distance between one corner
of a room and the farthest corner from it is $l^2 + w^2 + h^2$, where
l, w, and h are the length, width, and height of the room.
Computations based on the assumption of Euclidean distance treat
the dependent variables as independent spatial dimensions, like
length, width, and height of a room.

   First we will define F for MANOVA in terms of squared distances in multidimensional space and then show that these
squared distances can be determined without direct measurement.
This will be done only for a multivariate generalization of one-
way ANOVA, but analogous geometrical generalizations of repeated-measures and factorial ANOVA also are possible.

   For standardizing the dependent variables, to ensure that
scale of measurement alterations do not affect a MANOVA test,

each of the dependent variables is transformed to z scores on the basis of the mean and standard deviation of measurements of the dependent variable. The points involved in the following conception of MANOVA are, then, vectors of z scores.

Unlike the procedure of reducing a number of measurements on an experimental unit to a single number by the derivation of composite measurements, the geometrical approach reduces a number of measurements on an experimental unit to a single point in space. The distances between points in p-dimensional space are all that are involved in this form of MANOVA; the actual numerical values determining the location of a point are not required for the geometrical conception of the determination of F for MANOVA. If we regard a centroid as a multidimensional mean, F for one-way MANOVA can be defined in the same way as for univariate one-way ANOVA:

$$F = \frac{SS_B / df_B}{SS_W / df_W}$$

where sums of squares are sums of squared deviations from means (which are centroids in the multidimensional case), df_B is the number of treatment groups minus one, and df_W is the number of subjects (or other experimental units) minus the number of treatment groups. For MANOVA, let C_A, C_B, and C_{GR} be the centroids of groups A and B and of all of the vectors combined ("the grand centroid"). Then we can define SS_B and SS_W for MANOVA as:

$$SS_B = n_A (C_A - C_{GR})^2 + n_B (C_B - C_{GR})^2 + \cdots$$

$$SS_W = \Sigma (X_A - C_A)^2 + \Sigma (X_B - C_B)^2 + \cdots$$

The differences enclosed in parentheses for SS_B are straight-line distances between centroids, and the differences in parentheses for SS_W are straight-line distances between measurement vectors and centroids. The following formulas determine those squared distances by adding the squared deviations from means on each of the dependent variables (X, Y, etc.):

$$SS_B = SS_{B(X)} + SS_{B(Y)} + \cdots$$

$$SS_W = SS_{W(X)} + SS_{W(Y)} + \cdots$$

So for computational purposes, F for one-way MANOVA becomes:

$$\frac{(SS_{B(X)} + SS_{B(Y)} + \cdots)}{(SS_{W(X)} + SS_{W(Y)} + \cdots)} \times \frac{df_W}{df_B}$$

To obtain a simpler test statistic that is equivalent for the determination of significance by the randomization test procedure, we can drop the constant multiplier df_W/df_B. The numerator *plus* the denominator of the remaining fraction equals the sum of the total sum of squares for all of the dependent variables: $SS_{T(X)} + SS_{T(Y)} + \cdots$, which is constant over all data permutations. Therefore, the numerator varies inversely with the denominator over all data permutations, making the numerator an equivalent test statistic to the fraction as a whole. The numerator, in turn, can be reduced to a simpler test statistic. SS_B for each dependent variable is $n(\bar{X} - \bar{X}_{GR})^2$, and as the grand mean for each dependent variable in z score form is 0, the sum of SS_B for all dependent variables is $\Sigma n(\bar{X})^2$, or $\Sigma(T^2/n)$, where the summation is over all dependent variables and all groups.

7.7   EVALUATION OF THREE MANOVA TESTS

Three randomization test procedures for MANOVA have been described in this chapter. One used composite z scores, another involved the summing of F's over the dependent variables, and the third was a geometrical approach. Although attention was focused on one-way MANOVA, all three approaches can be adapted to repeated-measures or factorial MANOVA. The following evaluation concerns the tests as one-way MANOVA, the principal application for which they were described in this chapter.

All three techniques apparently require less time for the determination of significance by a randomization test than would a randomization test with a conventional MANOVA test statistic. The composite z score approach would be the fastest of the three for a randomization test, the time required after the composite z scores have been computed being the time for univariate ANOVA with observations on only one of the dependent variables.

The composite z score procedure has the added advantage of allowing the use of computer programs in preceding chapters as soon as the composite scores have been derived. The other two approaches would require new programs to be written.

A serious limitation of the geometrical approach is its inapplicability to one-tailed t tests. The geometrical approach permits spatial analogues of sums of squared deviations to be formed but not spatial analogues of the arithmetic difference between means that is the numerator of a one-tailed t. The sum of the squared differences between the means of A and B groups with respect to each dependent variable completely determines the squared distance (and the distance) between the A and B centroids in Euclidean space, but the sum of $(\bar{X}_A - \bar{X}_B)$ over all dependent variables does not determine the distance between the A and B centroids nor their relative positions in space. For instance, when the sum of the $(\bar{X}_A - \bar{X}_B)$ differences is 0, the A and B centroids are not necessarily located at the same point. Univariate sums of squares generalize readily in a geometric manner to multivariate situations but arithmetic differences between means do not.

A major strength of the composite z score procedure is its sensitivity to treatment differences when the way in which treatments covary can be predicted. The manipulation of signs of z scores of a variable to reflect the expected direction of correlation between the variables provides the opportunity to increase the likelihood of detecting treatment effects when more than two treatments are compared, but it is in the application of the t test that the chance to increase the power of the test is clearest. Consider the use of the total of the z scores for one of the treatments as the one-tailed test statistic. The one-tailed test can be based on any of the following predictions:

1. X and Y values will be positively correlated over treatments.
2. X and Y values will be negatively correlated over treatments.
3. Both X and Y values will be large for treatment A.
4. X values will be large and Y values will be small for treatment A.

The first two predictions are the most general. For prediction 1, the absolute value of the sum of group A is the appropriate test statistic. When prediction 2 is made, the sign of the z scores is reversed for either of the dependent variables, and the absolute value of the sum of A is the test statistic to use. The sum of A is the test statistic to use for prediction 3, and the same test statistic is used for prediction 4 after the signs of the z scores for Y are reversed. With more than two dependent variables, the predictions can be even more specific.

The test procedure based on the sum of test statistics can use the sum of one-tailed t's for the dependent variables for group A to test predictions 3 and 4. To test prediction 1, t's with $(\bar{X}_A - \bar{X}_B)$ and $(\bar{Y}_A - \bar{Y}_B)$ as numerators are added, and the absolute sum is used as the test statistic. To test prediction 2, t's with $(\bar{X}_A - \bar{X}_B)$ and $(\bar{Y}_B - \bar{Y}_A)$ are added and the absolute sum is used as the test statistic.

The only tests that cannot be carried out satisfactorily with the composite z score approach are those involving two or more groups where no assumption about the relationship between dependent variables is made. The adjustment of z scores to reflect the way in which the dependent variables are expected to covary is both a strength and a weakness of the procedure. Clearly, it is a strength when there is a good basis for prediction of the nature of covariation of dependent variable values over the treatments, but to use the procedure without such a prediction is inappropriate. Failing to change the sign of z scores for any of the dependent variables is implicitly to predict that all dependent variables change in the same direction from treatment to treatment, and the P-value is affected by the correctness of the implicit prediction. To carry out MANOVA or multivariate t tests, then, without expectations about the way the dependent variables covary, one should employ either the sum of statistics procedure or the geometrical procedure.

7.8 MULTIVARIATE FACTORIAL DESIGNS

A multivariate factorial test can be used for testing treatment effects in experiments with more than one independent and more than one dependent variable. In Example 6.12, a factorial experiment with two factors and two dependent variables was described, but each dependent variable was associated with only one of the factors. In the following example, we will consider the same experimental design, analyzed as it would be if both dependent variables were regarded as measures of effect for both factors.

Example 7.4

A physiological psychologist is interested in the effect of minor lesions in the brains of rats on the ability of the rats to discriminate among intensities of stimuli. In addition to investigating the relative performance of lesioned and nonlesioned rats, the experimenter also wants to compare the effect of lesions in different locations in the brain. Brightness and loudness discrimination tasks are convenient ways to assess deterioration of discrimination ability, and the experimenter decides to measure both brightness and loudness discrimination in order to have a more powerful assessment of lesion effects. To make effective use of a small number of rats, the experimenter decides to use a factorial design in which rats are assigned to levels of two factors, and measurements of brightness and loudness discrimination are made on all rats. Five rats are assigned at random to each of the four treatment combinations in the following matrix:

		Lesion in location L_1?	
		Yes	No
Lesion in	Yes	n = 5	n = 5
location L_2?	No	n = 5	n = 5

Five rats have no lesions, five have lesions in both L_1 and L_2, five have a lesion only in L_1, and five have a lesion only in L_2.

After all of the lesioned rats have recovered from the operation, all lesioned and unlesioned rats are tested on brightness and loudness discrimination tasks. Each rat provides a measurement on two dependent variables: brightness discrimination and loudness discrimination.

The experimenter expects brightness and loudness discrimination both to be adversely affected by the presence of a lesion, and for all of the t tests he uses the composite z score approach to determine significance. The 20 pairs of measurements are transformed into 20 composite z scores, one for each rat. The 20 composite z scores constitute the data to be permuted to test for treatment effects.

Various tests are possible. To see whether the presence of a lesion reduces ability to discriminate intensities of stimuli, the composite z scores for the five animals receiving no lesions plus the composite z scores for the five animals receiving both lesions could be compared by the use of Program 4.4, a randomization test program for the independent t test, using the total for the five animals with no lesions as the test statistic. (If lesioning had an adverse effect, then with the largeness of the z scores reflecting the ability to discriminate, the nonlesioned rats should have larger z scores.) Program 6.1 could be used to test the main effect of the lesion in L_1 by permuting the data within rows, or to test the effect of the lesion in L_2 by permuting within columns. The relative effect of lesions in L_1 and L_2 could be tested by use of Program 4.4 applied to the ten composite z scores of the rats that had only one lesion, based on $10!/5!5!$ permutations of the values for those rats.

7.9 REPORTING RESULTS OF STATISTICAL TESTS

If a conventional multivariate ANOVA test statistic is used, with significance determined by a randomization test, the research report should indicate which of the several conventional test statistics was used. The other tests described in this chapter are new and require description as well as designation as randomization tests. The method of computing test statistics for these tests should be described in detail.

REFERENCES

Harris, R. J. (1975). *A Primer of Multivariate Statistics*. Academic Press, New York.

Olson, C. L. (1976). On choosing a test statistic in multivariate analysis of variance. *Psychol. Bull. 83*, 579—586.

8
Correlation

Significance tables for the product-moment correlation coefficient assume random sampling of a bivariate normal distribution and, therefore, would not be appropriate for experiments even if the subjects were randomly selected; also there would have to be random selection of the independent variable values from a normal distribution of potential values, and such a requirement is unrealistic.

8.1   DETERMINING SIGNIFICANCE BY DATA PERMUTATION

Nevertheless, correlation would seem to be a useful procedure to employ for detecting experimental treatment effects, and the problem of validity can be overcome by determining significance by the use of data permutation. The correlation between a quantitative treatment variable and measurements can constitute a more powerful test than the division of subjects into two or three treatment groups. For example, if we have only five subjects and one measurement from each subject, with random assignment to two treatments the smallest P-value that can be obtained by a randomization test is 0.10, because there are only 10 data permutations associated with the sample sizes providing the most data permutations: two subjects for one treatment and three for the other. On the other hand, if five treatments consisting of

five magnitudes of the independent variable were used instead
of two treatments, with random assignment of one subject per
treatment, the P-value for the correlation between the independent variable values and the response measurements could be
small. The P-value could be as low as 1/120 because there are
5! = 120 data permutations associated with the 5! ways five subjects can be assigned to five treatments, with one subject per
treatment. Thus an experimenter can have a relatively sensitive
experiment with very few subjects, provided that he employs a
number of degrees of magnitude of treatment and uses a correlation test statistic.

Example 8.1

We want to test the effect of a certain drug on the reaction time
of subjects. One way would be to compare the drug with a
placebo and to use the t test. This is somewhat like comparing
the effect of zero amount of a drug with X amount. To test
whether three different amounts of a drug have the same effect
is not the same as testing whether presence or absence of a
drug has the same effect, but sometimes it is just as important.
We take three drug dosages, consisting of 5, 10, and 15 units
of the drug and randomly assign three subjects to the dosages,
with one subject per dosage. To illustrate how the significance
of the treatment effect is determined by a randomization test, we
will use the following hypothesized results and assume that a
positive correlation was·predicted:

Drug dosage	5	10	15
Response	20 (1)	22 (2)	25 (3)

The index numbers are assigned in sequential order to the
response measurements and are shown in parentheses following
the measurements. A standard formula for the product-moment
correlation coefficient r, formula (8.1) given in Section 8.2, is
applied to the results and r is found to be 0.99. The significance of r based on data permutation is determined from the six
possible assignments of subjects to treatments. Given that the
assignment had no effect on the measurement of any subject, the
six possible assignments would provide the six data permutations
in Table 1, where the numbers in parentheses are indexes
which determine the listing of the data permutations.

TABLE 1 List of Data Permutations

		Permutation 1	
Drug dosage:	5	10	15
Response:	20 (1)	22 (2)	25 (3)
	r = 0.99		
		Permutation 2	
Drug dosage:	5	10	15
Response:	20 (1)	25 (3)	22 (2)
	r = 0.40		
		Permutation 3	
Drug dosage:	5	10	15
Response:	22 (2)	20 (1)	25 (3)
	r = 0.60		
		Permutation 4	
Drug dosage:	5	10	15
Response:	22 (2)	25 (3)	20 (1)
	r = −0.40		
		Permutation 5	
Drug dosage:	5	10	15
Response:	25 (3)	20 (1)	22 (2)
	r = −0.60		
		Permutation 6	
Drug dosage:	5	10	15
Response:	25 (3)	22 (2)	20 (1)
	r = −0.99		

The indexes for each permutation can be regarded as form-
ing a three-digit number which increases in magnitude from 123
for the first permutation to 321 for the sixth permutation. For
correlation, the index numbers 1 to N are assigned to the N
measurements in the order in which they are associated with the
ordered treatments for the obtained results. The listing procedure, then, is to have the sequence of indexes for each permutation treated as an N-digit number and to permute the sequence of index numbers in such a way as to make each successive N-digit number the next larger value that can be
obtained by permuting the order.

If the prediction was a strong positive correlation, not just
a strong correlation, between the dosage and response magnitudes, the significance of the obtained r is the proportion of
the six permutations that provide such a large value of r as the
obtained value, 0.99. Only the first permutation, which represents the obtained results, shows such a large r, and so the
P-value for the obtained results is 1/6 or about 0.167.

If it had been anticipated that the correlation between drug
dosage and response would be strong, but the direction of correlation was not predicted, the absolute value of the correlation
coefficient $|r|$ would be an appropriate test statistic. Permuta-
tions 1 and 6 both provide values of $|r|$ as large as the obtain-
ed value, and so the two-tailed significance value would be 2/6,
or about 0.333.

For any situation where, as in Example 8.1, the independent
variable values are uniformly spaced or are otherwise symmetrically distributed about the middle of the distribution (e. g., 5,
20, 40, 55), the two-tailed P-value for r will be twice the one-
tailed probability for a correctly predicted direction of correlation.
That is not necessarily the case, however, when the independent
variable values are not symmetrically distributed.

Example 8.2

Suppose that the highest drug dosage had been 20 units and
that the same responses had been obtained. Then the first data
permutation would have been:

Drug dosage	5	10	20
Response	20 (1)	22 (2)	25 (3)
		r = 0.997	

This would be the data permutation providing the largest positive correlation coefficient, so that if a positive correlation was predicted the significance value would be 1/6, or about 0.167. The two-tailed P-value, however, would not be double that value but would be the same value, because the first data permutation provided the largest $|r|$ value; the largest negative r value would occur for the sixth data permutation, where the sequence of responses was reversed, and it would have a smaller absolute value ($r = -0.95$) than the obtained positive correlation.

Thus, for asymmetrical distributions of the independent variable, the two-tailed P-value need not be twice the one-tailed P-value, and so the one-tailed and two-tailed P-values must be derived separately.

8.2 COMPUTER PROGRAM FOR SYSTEMATIC DATA PERMUTATION

In using a randomization test for determining significance by either systematic or random data permutation, a simpler but equivalent test statistic to r can be used. A common formula for determining r by use of a desk calculator is:

$$r = \frac{N\Sigma XY - \Sigma X \Sigma Y}{\{[N\Sigma X^2 - (\Sigma X)^2][N\Sigma Y^2 - (\Sigma Y)^2]\}^{1/2}} \tag{8.1}$$

It gives the same value of r as other formulas, but it is especially useful for deriving simpler equivalent test statistics for data permutation.

Over the data permutations for correlation, the series of X values is paired with the series of Y values in every possible way. The denominator of formula (8.1) is unaffected by the way in which the values are paired, and so it is a constant over all data permutations. As a constant divisor it can be eliminated, leaving the numerator as an equivalent test statistic to r. The second term of the numerator also is constant over all data permutations, and so it too can be eliminated, leaving $N\Sigma XY$ as an equivalent test statistic to r. As N is a constant multiplier it too can be eliminated, leaving ΣXY as an equivalent test statistic to r. We have found, then, that the proportion of data permutations providing as large a value of ΣXY as an obtained value is the proportion providing as large a value of r as the obtained value. The minimum possible ΣXY for a set of data, therefore, is associated with the strongest negative correlation, and the maximum possible ΣXY is associated with the strongest positive correlation. Thus, to find the significance value for a one-tailed test, we find the proportion of data permutations providing a value of ΣXY as large as the obtained value for a predicted positive correlation and the proportion with a ΣXY as small as the obtained value for a predicted negative correlation.

The absolute value $|r|$ is the test statistic for a two-tailed test, where the direction of correlation has not been predicted. Since the denominator of formula (8.1) remains constant over all data permutations, the absolute value of the numerator is an equivalent test statistic to $|r|$. Thus the two-tailed test statistic used in the computer programs that follow is $|N\Sigma XY - \Sigma X\Sigma Y|$.

Now we will consider the principal operations that the computer must perform to determine the significance by systematic data permutation, when Program 8.1 is used.

Step 1. Arrange the N independent variable (X) values in order from low to high and allocate index numbers 1 to N to the measurement (Y) values according to the independent variable values with which they are associated, 1 being assigned to the measurement for the lowest independent variable value and N being assigned to the measurement for the highest independent variable value.

Step 2. Use the procedure described in Section 8.1 for generating a systematic sequence of permutations of the index numbers of the Y values, the first sequence of index numbers being 1, 2, . . . , N − 1, N, representing the obtained results, the second sequence being 1, 2, . . . , N, N − 1, and so on, the last sequence being N, N − 1, . . . , 2, 1. The corresponding measurements for each of the sequences are paired with the X values for the same sequential position, and steps 3 and 4 are performed for each sequence. The first sequence of index numbers represents the obtained results, and so the performance of steps 3 and 4 on the first sequence gives the obtained test statistic value.

Step 3. Compute ΣXY, the sum of the products of the paired X and Y values, as the one-tailed test statistic.

Step 4. Compute $|N\Sigma XY - \Sigma X\Sigma Y|$ as the two-tailed test statistic.

Step 5. For a one-tailed test where a positive correlation is predicted, compute the one-tailed probability value as the proportion of the N! permutations, including the one representing the obtained results, that provide a value of ΣXY greater than or equal to the obtained ΣXY. For a one-tailed

PROGRAM 8.1 Product-moment
Correlation: Systematic
Permutation

```
          ((INPUT))

   READ(5,---)N
   OBTONE=OBTTWO=TOTX=TOTY=0
   DO 1 I=1,N
   READ(5,---)X(I),Y(I)
   OBTONE=OBTONE+X(I)*Y(I)
   TOTX=TOTX + X(I)
   TOTY=TOTY + Y(I)
   INDEX (I)=I
 1 WOR(I)=1
   PRODTOT=(TOTX)*(TOTY)
   OBTTWO=ABS((N*OBTONE)-PRODTOT)
   NGEONE=NGETWO=NPERM=1
 2 WOR(INDEX(N))=0
   I=N-1
 3 WOR(INDEX(I))=0
 4 IF(INDEX(I).EQ.N)GO TO 6
   INDEX (I)=INDEX(I)+1
 5 IF(WOR(INDEX(I)).EQ.1)GO TO 4
   WOR(INDEX(I))=1
   IF(I.EQ.N)GO TO 7
   I=I+1
   INDEX (I)=1
   GO TO 5
 6 IF(I.EQ.1)GO TO 9
   I=I-1
   GO TO 3
 7 TESTONE=TESTTWO=0
   NPERM=NPERM + 1
   DO 8 I=1,N
 8 TESTONE=TESTONE+X(I)*Y(INDEX(I))
   TESTTWO=ABS((N*TESTONE)-PRODTOT)
   IF(TESTONE.GE.OBTONE)NGEONE=NGEONE+1
   IF(TESTTWO.GE.OBTTWO)NGETWO=NGETWO+1
   GO TO 2
 9 PROBONE=FLOAT(NGEONE)/NPERM
   PROBTWO=FLOAT(NGETWO)/NPERM

          ((OUTPUT))
```

---

test where a negative correlation is predicted, the P-value
is the proportion of the permutations with a value of ΣXY
less than or equal to the obtained ΣXY.

Step 6. For a two-tailed test, the P-value is the proportion
of the permutations that provide a value of $|N\Sigma XY - \Sigma X\Sigma Y|$
greater than or equal to the obtained $|N\Sigma XY - \Sigma X\Sigma Y|$.

Program 8.1 goes through these steps in determining the significance of r by systematic data permutation. The P-value,
however, is for a one-tailed test where a positive correlation is
predicted. If a negative correlation is predicted, a single change
in a statement in the program makes it appropriate: Change the
statement "IF(TESTONE.GE.OBTONE)NGEONE = NGEONE + 1"
to the following statement: "IF(TESTONE.LE.OBTONE)NGEONE =
NGEONE + 1."

Program 8.1 can be tested with the following data, for which
the one-tailed P-value for a predicted positive correlation is
2/120, or about 0.0167, and the two-tailed P-value is 4/120, or
about 0.0333:

X	2	4	6	8	10
Y	1	3	5	8	7

8.3   CORRELATION WITH RANDOM DATA
PERMUTATION

The principal operations that the computer must perform to determine the significance of the product-moment correlation coefficient
by random data permutation with Program 8.2 are as follows:

Step 1. Arrange the N independent variable (X) values in order
from low to high and allocate index numbers 1 to N to the
measurement (Y) values according to the independent variable values with which they are associated, 1 being assigned to the measurement for the lowest independent variable
value and N being assigned to the measurement for the
highest independent variable value.

Step 2. Compute ΣXY, the sum of the product of the paired X
and Y values, as the obtained one-tailed test statistic.

Step 3. Compute $|N\Sigma XY - \Sigma X\Sigma Y|$ as the obtained two-tailed
test statistic.

PROGRAM 8.2   Product-moment
Correlation: Random Permutation

((INPUT))

```
    READ (5,---)N
    Z=RANSET(12.34)
    NGEONE=NGETWO=TOTX=TOTY=0
    DO 1 I=1,N
    READ (5,---)X(I),Y(I)
    TOTX=TOTX+X(I)
  1 TOTY=TOTY+Y(I)
    PRODTOT=(TOTX)*(TOTY)
    READ (5,---)NPERM
    DO 3 J=1,NPERM
    TESTONE=TESTTWO=0
    DO 2 I=1,N
    IF(J.EQ.1)GO TO 2
    ID=I+INT(RANF(0)*(N-I+1))
    Z=X(I)
    X(I)=X(ID)
    X(ID)=Z
  2 TESTONE=TESTONE+X(I)*Y(I)
    TESTTWO=ABS((N*TESTONE)-PRODTOT)
    IF(J.EQ.1)OBTONE=TESTONE
    IF(J.EQ.1)OBTTWO=TESTTWO
    IF(TESTONE.GE.OBTONE)NGEONE=NGEONE+1
  3 IF(TESTTWO.GE.OBTTWO)NGETWO=NGETWO+1
    PROBONE=FLOAT(NGEONE)/NPERM
    PROBTWO=FLOAT(NGETWO)/NPERM
```

((OUTPUT))

Step 4.   Use a random number generation algorithm that will
    provide a permutation of the index numbers 1 to N, that is,
    an algorithm that will randomly order the index numbers 1
    to N.   Where NPERM is the requested number of data per-
    mutations, perform NPERM − 1 data permutations and com-
    pute ΣXY and $|N\Sigma XY - \Sigma X\Sigma Y|$ for each.
Step 5.   For a one-tailed test where a positive correlation is pre-
    dicted, the significance of r is the proportion of the NPERM
    data permutations (the NPERM − 1 generated by the random
    number generation algorithm plus the obtained results) that
    provide a value of ΣXY greater than or equal to the obtained
    value.   Where a negative correlation is predicted, the

significance is the proportion that provide a value of ΣXY
less than or equal to the obtained value.

Step 6. For a two-tailed test, the significance of r is the proportion of the data permutations that provide a value of the
test statistic $|N\Sigma XY - \Sigma X\Sigma Y|$ greater than or equal to the
obtained value.

Program 8.2 goes through these steps in determining the
significance of r by random data permutation. As with Program
8.1, the P-value given for a one-tailed test is for a prediction
of a positive correlation. If a negative correlation is predicted,
the statement "IF(TESTONE.GE.OBTONE)NGEONE = NGEONE + 1"
should be replaced by the statement "IF(TESTONE.LE.OBTONE)
NGEONE = NGEONE + 1" so that the P-value becomes the proportion of the data permutations with a value of ΣXY *as small
as* the obtained value.

Program 8.2 can be tested with the following data, for which
systematic data permutation would give a one-tailed P-value of
2/120 or about 0.0167, for a predicted positive correlation and
a two-tailed P-value of 4/120, or about 0.0333:

X	2	4	6	8	10
Y	1	3	5	8	7

8.4   MULTIVARIATE CORRELATION

Multivariate correlation for experiments is that for which there
are two or more dependent variables. We will consider two types
of situations: those situations in which all of the dependent
variables are expected to be correlated with the treatment values
in the same way, that is, all positively or all negatively correlated, and those situations in which the direction of correlation
may vary over the dependent variables. For both types of multivariate correlation the first step is to transform the values for
each dependent variable into z scores.

When all dependent variables are expected to show the same
direction of correlation with the independent (treatment) variable, a composite z score, which is the sum of the z scores for
the dependent variables, is used as a composite measurement
for a subject for a particular independent variable value. (See
Section 7.4 regarding construction of composite measurements.)
The N! data permutations used to determine significance consist

of every possible pairing of the N composite measurements with the N independent variable values, and for each data permutation a one- or two-tailed test statistic is computed.

When the nature of the variation in the direction of correlation between dependent variables and the independent variable is unpredictable, it is not appropriate to add the z scores for each subject. Instead, the one-tailed test statistic ΣXY or the two-tailed test statistic $|N\Sigma XY - \Sigma X\Sigma Y|$ should be computed for the z scores for each dependent variable separately. For each data permutation, the multivariate test statistic is the sum of the univariate test statistics. For example, for a one-tailed test, the multivariate test statistic is ΣXY for the first dependent variable plus ΣXY for the second, and so on, where the Y values are z scores.

8.5 POINT-BISERIAL CORRELATION

A point-biserial correlation coefficient is a product-moment correlation coefficient based on the correlation between a dichotomous and a continuous variable. For experimental applications, the dichotomous variable can be either the independent or the dependent variable.

When the independent variable for a completely randomized experiment is dichotomous and the dependent variable is continuous, Program 4.4 gives P-values for t and $|t|$ that are the same as would be given by Program 8.1 for r and $|r|$, using 0's and 1's as the X values to be paired with the continuous Y values. The two programs give the same P-values for this kind of application because, as will be shown below, the test statistics for Program 4.4, which are T_L and $\Sigma(T^2/n)$, are equivalent to the test statistics for Program 8.1, which are ΣXY and $|N\Sigma XY - \Sigma X\Sigma Y|$.

To see the equivalence of the test statistics in the two programs, consider the following X and Y values (page 208) for Program 8.1, where a 1 indicates treatment A (the treatment predicted to provide the larger measurements) and a 0 indicates treatment B (predicted to provide the smaller measurements). $\Sigma X = n_A$ because the only nonzero X values are the 1's denoting applications of treatment A. Each Y value associated with treatment A yields an XY value that is the same as the Y value, and each Y value associated with treatment B yields an XY value of 0; thus, $\Sigma XY = T_A$, the total of the measurements for treatment A.

X	Y	XY
1	Y_1	Y_1
1	Y_2	Y_2
.	.	.
.	.	.
.	.	.
0	Y_{N-1}	0
0	Y_N	0
$\Sigma = n_A$	ΣY	T_A

The numerical identity of ΣXY and T_A shows the equivalence of the one-tailed test statistics r and t. The equivalence of the two-tailed test statistics $|r|$ and $|t|$ can be determined by examining a formula (Nunnally, 1967, p. 120) for point-biserial r:

$$r_{pb} = \frac{\overline{X}_A - \overline{X}_B}{\sigma}\sqrt{pq} \qquad (8.2)$$

where $\overline{X}_A$ and $\overline{X}_B$ are the means of the continuous variable associated with the two levels of the dichotomous variable, σ is the standard deviation of the continuous variable for both groups combined, and p and q are proportions of subjects for the two levels of the dichotomous variable. Over all data permutations, only $\overline{X}_A - \overline{X}_B$ varies, so $|r|$ is an equivalent test statistic to $|\overline{X}_A - \overline{X}_B|$ and thus to $|t|$. In order to understand the following demonstration of the equivalence of $|N\Sigma XY - \Sigma X\Sigma Y|$ and $|\overline{X}_A - \overline{X}_B|$ it will be necessary to keep in mind that $\Sigma X = n_A$ and $\Sigma XY = T_A$.

The two-tailed correlation test statistic in Program 8.1, $|N\Sigma XY - \Sigma X\Sigma Y|$ can be expanded to $|(n_A + n_B)(n_A\overline{X}_A) - (n_A)(n_A\overline{X}_A + n_B\overline{X}_B)|$, which is algebraically equal to $|(n_A n_B)\overline{X}_A - \overline{X}_B)|$. Since $n_A n_B$ is a constant multiplier over all data permutations, it can be dropped, leaving $|\overline{X}_A - \overline{X}_B|$ and thus $|t|$ as an equivalent test statistic to $|r|$.

Program 8.1 is not as useful as Program 4.4 for point-biserial correlation because it performs ($n_A!n_B!$) times as many data permutations as Program 4.4 in order to get the same P-values. This redundancy arises from the fact that for *each division of Y values between A and B* (treatments represented by 1 and 0), there are $n_A!$ pairings of the values for A with the 1's in the X column and $n_B!$ pairings of the values for B with the 0's in the X column.

Next, consider point-biserial correlation when we have a continuous *independent* variable and a dichotomous *dependent* variable. The output of Program 8.1 is unaffected by which variable is called the X variable and which the Y variable, even though the order in which members of a pair of X and Y values are entered does affect the way the data are permuted, because the sequential order of values for one variable are held fixed while the sequence of values for the other variable is permuted. Which variable is called X and has its sequence fixed while the other variable's sequence is permuted does not affect the distribution of ($n_A + n_B$)! pairings and so we can, as in the preceding proof of the equivalence of the test statistics for Programs 4.4 and 8.1, list the continuous variable under Y and the dichotomous variable under X. The preceding demonstration of equivalence of the test statistics for the two programs now can be seen to apply also to situations where the independent variable is continuous and the dependent variable is dichotomous. So, whether the independent variable is dichotomous and the dependent variable continuous or vice versa, Program 4.4 can be used instead of Program 8.1. Furthermore, since Program 4.4 performs the same test with fewer data permutations, it is the preferred program. In using Program 4.4 when the independent variable is continuous, the values of the independent variable are treated as if they were the data, and the dichotomous results of the study are treated as if they represented two treatments.

Example 8.3

An experimenter assigns five dosages of a drug randomly to five animals, with one dosage per animal, and records the presence (1) or absence (0) of adverse side effects. The following results are obtained:

X:	5	10	15	20	25
Y:	0	0	1	1	1

To statistically test the effect of dosage level, the experimenter uses Program 4.4 as if the five animals had been assigned randomly to treatments A and B, with three animals for A (designated as treatment 1) and two animals for B (designated as treatment 0). Within this framework, the obtained "data" for treatment A are 15, 20, and 25, and the obtained "data" for treatment B are 5 and 10. These five numbers are divided in $5!/2!3! = 10$ ways by Program 4.4 to determine the one-tailed and two-tailed P-values.

When the continuous variable consists of ranks, Program 4.4 still can be used, of course, whether the ranks are ranks of treatments or ranks of data. However, an alternative procedure would be to apply the Mann-Whitney U test to the ranks and use published tables to get the same P-values as those provided by Program 4.4.

8.6 CORRELATION BETWEEN DICHOTOMOUS VARIABLES

Next, consider applications of correlation to experiments with dichotomous independent and dependent variables. Application of Program 8.1 to the data, where the 0's and 1's representing the treatments are paired with 0's and 1's representing the experimental results, gives the same significance for r and $|r|$ as Program 4.4 gives for t and $|t|$ for the same data. (The equivalence of the test statistics for the two programs described in Section 8.5 obviously holds when both variables are dichotomous as well as when one is dichotomous and the other continuous.)

As discussed in Section 4.13, the P-value for t given by Program 4.4 is that for Fisher's exact test for a 2×2 contingency table, and the P-value for $|t|$ given by Program 4.4 is the P-value for contingency chi-square when both the independent and the dependent variables are dichotomous. Thus, the application of Program 4.4 or 8.1 to dichotomous data from two treatments provides the P-value for Fisher's exact test (one-tailed test) and the P-value for the contingency chi-square test (two-tailed test).

8.7 SPEARMAN'S RANK CORRELATION PROCEDURE

Spearman's rank correlation procedure is a commonly used nonparametric procedure for which published significance tables are

available. The Spearman rank correlation coefficient R is a product-moment correlation coefficient where the correlation is based
on ranks instead of raw measurements. The independent and
dependent variable values are ranked separately from 1 to N,
and R is computed from the paired ranks by the following
formula:

$$R = 1 - \frac{6\Sigma D^2}{N(N^2 - 1)} \tag{8.2}$$

where D is the absolute difference between a pair of ranks and
N is the number of pairs of ranks. When data for both X and
Y variables are expressed as ranks with no tied ranks, the
P-value given by Program 8.1, for systematic data permutation,
is the same as the value given by the significance table for
Spearman's R, since the table is based on data permutation.
For experimental situations where both the independent and de-
pendent variables are expressed as ranks, it is therefore un-
necessary to determine significance by data permutation because
the significance tables for R provide the data permutation
P-values. (Of course, the tables frequently are restricted to
$\alpha = 0.05$ and $\alpha = 0.01$.)

Gibbons (1971, pp. 228–232) showed that, over the data
permutations used for determining the significance of R,

$$E(R) = 0 \tag{8.3}$$

$$\sigma^2(R) = \frac{1}{N - 1} \tag{8.4}$$

The data permutation distribution of R approaches normality
rapidly with an increase in N, the number of pairs of ranks.
Consequently, for relatively small N a close approximation to the
P-value given by systematic data permutation can be provided by
the use of normal curve tables, on the basis of formulas (8.3)
and (8.4). To use the normal curve tables, R is regarded as
normally distributed, with

$$\mu = 0 \tag{8.5}$$

$$\sigma = \left(\frac{1}{N - 1}\right)^{1/2} \tag{8.6}$$

The normal curve approximation is very close even when N is as
small as 14. Tables of R show that values of 0.456 and 0.645 are

required for one-tailed significance at the 0.05 and 0.01 levels, for N = 14. Formulas (8.5) and (8.6) provide z scores of 1.64 and 2.33 for R values of 0.456 and 0.645, and reference to normal curve tables gives significance values of 0.0505 and 0.0099, which certainly are close to 0.05 and 0.01.

8.8   REPORTING RESULTS OF STATISTICAL TESTS

The two computer programs in this chapter are for product-moment correlation, with significance determined by the randomization test procedure. The product-moment correlation coefficient for the experimental results could be reported along with the associated degrees of freedom and the P-values, with the statement that the P-value was based on a randomization test. Spearman's rank correlation coefficient would be reported with the associated degrees of freedom and the P-value given in the significance tables or the P-value based on a randomization test, when appropriate.

When the independent and dependent variables both are dichotomous, report the test as being applied to a 2 × 2 contingency table. The one-tailed P-value is for Fisher's exact test and the two-tailed P-value is for contingency chi-square and should be reported along with the contingency chi-square value for the obtained results. When the independent variable is dichotomous and the dependent variable is continuous, the test should be described as a t test with significance determined by data permutation, and the obtained value of t and the associated degrees of freedom should be given in the report. When the independent variable is continuous and the dependent variable is dichotomous, a point-biserial correlation coefficient should be computed by applying the product-moment correlation procedure to the paired continuous and dichotomous values, and this correlation coefficient and degrees of freedom should be reported. The P-value given by Program 8.1 or Program 4.4 is the significance of the point-biserial correlation coefficient based on data permutation.

Other tests in this chapter are unconventional and when they are used the reporting of the results is more complex. The multivariate correlation tests are tests which require some explanation in the research report in order to provide the reader with a general understanding of the data analysis. The rationale for the use of z scores for multivariate correlation, which is to give equal weight to the dependent variables, regardless of the units of measurement, should be given when multivariate correlation with composite measurements is described. (The discussion of composite z scores in Section 7.4 has a bearing on the use of z scores for multivariate correlation.)

REFERENCES

Gibbons, J. D. (1971). *Nonparametric Statistical Inference.* McGraw-Hill, New York.

Nunnally, J. C. (1967). *Psychometric Theory.* McGraw-Hill, New York.

9
Trend Tests

The principal function served by parametric trend analysis is to determine the relative contribution of linear, quadratic, and higher-order polynomial functions to the variation of measurements over a quantitatively ordered series of treatments. The trend tests in this chapter have a different function, which is to test H_0 of no treatment effect. When the experimenter has reason to expect a particular kind of trend in the data over various magnitudes of the independent variable, trend tests are more sensitive than tests that do not predict a particular trend.

9.1 GOODNESS-OF-FIT TREND TEST

The goodness-of-fit trend test (Edgington, 1975) was developed to accommodate very specific predictions of experimental results for Price (Price and Cooper, 1975). Price replicated and extended work done by Huppert and Deutsch (1969) on the retention of an avoidance learning response as a function of the temporal length of the interval between training and testing. Price used training-testing intervals of various lengths, ranging from 30 minutes to 17 days, as the levels of the treatment variable. She wanted a test that would be quite powerful if the experimental data closely matched her prediction: small measurements for the 4-day training-testing interval and increasingly larger measurements for shorter and longer intervals. That is, she predicted a U-shaped trend with the low point of the U located at the 4-day training-testing interval. This prediction was based in part on the results of the study by Huppert and Deutsch, and in part on pilot work by Price. Huppert and Deutsch had obtained a U-shaped distribution of the mean number of trials taken by rats to relearn a task plotted against the length of the training-testing interval. Therefore, Price expected to get U-shaped distributions from her experiments. She used a different level of shock and expected that difference to alter the location of the low point of the U. Consequently, the low point was predicted on the basis of a pilot study by Price instead of on the results of Huppert and Deutsch.

Price had a specific prediction and wanted a procedure that would take into consideration its specificity. The first procedure considered was orthogonal polynomial trend analysis with significance determined in the usual way, by reference to F tables. There was the question, however, of whether the use of such a complex and little used technique would induce critical examination of the tenability of the parametric assumptions. Comparative and physiological psychology journals, which would be the appropriate media for Price's study, tend to use nonparametric tests rather frequently and this might reflect a critical attitude regarding parametric assumptions. (The use of one-way ANOVA, with significance determined by F tables, might go unquestioned by those journals but the test would not be sensitive to the predicted trend.) Consequently polynomial trend analysis was judged to be inappropriate. The second possibility considered was the use of parametric polynomial trend analysis with significance determined by a randomization test. Several difficulties arose in trying to do this.

First, the books at hand that discussed trend analysis restricted consideration to cases with equal sample sizes and equally spaced magnitudes of the independent variable, and neither of these restrictions was met by the experiment. Admittedly, there are journal articles that describe parametric trend analysis for unequal sample sizes and unequally spaced magnitudes of the independent variable, but researching such articles would have been worthwhile only if the trend analysis procedure was otherwise useful for the application.

Even superficial consideration of the standard trend analysis procedure, however, suggested its inadequacy for the desired analysis of the data. The standard test for quadratic trend, the test to use for prediction of a U-shaped data distribution, could not utilize the specificity of the prediction. The parametric trend test is two-tailed in the sense of not permitting differential predictions for U-shaped and inverted U-shaped trends, whereas

Price's prediction was not a general prediction of that kind but of a U-shaped trend in most cases. Second, the standard parametric trend procedure uses published tables of polynomial coefficients which do not permit specifying a predicted point of symmetry (low point for a U-shaped curve) near the lower end of the range of the independent variable magnitudes, and so the 4-day low-point prediction could not be used.

Because of these considerations, the use of standard parametric trend analysis with significance determined by means of a randomization test was ruled out in favor of developing a new test with a test statistic that could utilize highly specific predictions of trend. The randomization trend test that was developed for this purpose is called the *goodness-of-fit* trend test because of the similarity between its test statistic and that of the chi-square goodness-of-fit test. The goodness-of-fit trend test can be used with various types of predicted trend in addition to the U-shaped trend for which the test was originally developed.

9.2 POWER OF THE GOODNESS-OF-FIT TREND TEST

After developing the goodness-of-fit test, it was decided to compare the significance given by that test with that given by the alternative test that the experimenter would have used: one-way ANOVA with significance based on F tables. As mentioned earlier, the complex assumptions underlying orthogonal polynomial trend analysis made it unacceptable because of anticipated objections by journal editors. On the other hand, there probably would have been little objection to using simple one-way ANOVA with significance determined by F tables, and so it was a possible alternative to the goodness-of-fit test. Consequently, each set of data was analyzed by both the goodness-of-fit test and simple one-way ANOVA (with significance values from the F table) to permit a comparison of the significance given by the two procedures.

Example 9.1

First consider some training-testing interval data for which simple one-way ANOVA gave a P-value greater than 0.20, while the goodness-of-fit test for U-shaped trends gave a P-value of 0.007:

Time	30 min	1 day	3 days	5 days	7 days	10 days
Means	45.5	40.4	33.1	31.9	36.2	43.1

The data conform well to a U-shaped distribution with a low point at 4 days, and so the randomization test P-value is small.

Example 9.2

Next consider training-testing interval data where ANOVA gave a P-value between 0.05 and 0.10, while the goodness-of-fit test gave a P-value of 0.004:

Time	30 min	1 day	3 days	5 days	7 days
Means	12.6	9.9	6.2	6.4	9.6

Time	9 days	10 days	11 days	14 days	17 days
Means	13.6	14.5	13.7	14.1	18.7

Although the trend is not as consistent as in the previous example, the means do increase in value above and below the predicted low point of 4 days and the inconsistency in the upward trend from 10 to 14 days is so slight that the trend over the entire 17 days matches the predicted trend very closely. Therefore, the P-value for the randomization test is considerably lower than that for one-way ANOVA, where no predicted trend of any kind is involved.

Example 9.3

The final example based on Price's training-testing interval data is one where she predicted an *inverted* U-shaped trend. ANOVA provided a P-value of 0.01 whereas the goodness-of-fit test gave a much larger P-value, a value of approximately 1:

Time	30 min	1 day	3 days	10 days
Means	0.46	0.43	0.43	0.36

In this case, the randomization goodness-of-fit test gave a large P-value because the data deviate considerably from the predicted inverted U-shaped trend for this particular experiment in the study. The test determines the probability, under H_0, of getting a distribution of means that so closely matches the expected trend, and in this case the goodness-of-fit probability is about 1 because the obtained trend deviates so much from the predicted trend.

The goodness-of-fit test for trend is quite powerful when
the trend is rather accurately predicted but, as illustrated in
Example 9.3, it is very weak when the observed trend differs
considerably from the predicted trend. Therefore, the test is
very powerful only if it is used with discretion. As some critics
of one-tailed tests might say, a directional test should be used
only if any treatment differences in the opposite direction are
regarded as either irrational or irrelevant.

9.3   TEST STATISTIC FOR THE GOODNESS-OF-FIT TREND TEST

The goodness-of-fit trend test can be used to test the null hypothesis of no differential treatment effect by comparing experimental results with a predicted trend. The predicted trend is
represented by a distribution of "trend means" for the levels of
the treatment. These trend means are computed in various ways,
depending on the type of trend that is predicted. The goodness-
of-fit trend test statistic is $n_1(\bar{X}_1 - TM_1)^2 + \cdots + n_k(\bar{X}_k - TM_k)^2$, where n_k is the number of subjects in the kth treatment
group, $\bar{X}_k$ is the mean for the kth group, and TM_k is the trend
mean for the kth group. In other words, the test statistic is
the weighted sum of squared deviations of the measurement means
from the corresponding trend means. The closer the means to
the trend means, the smaller is the test statistic value, and the
P-value associated with the experimental results is the proportion
of the test statistic values in the reference set that are *as small
as* the obtained test statistic value.

9.4   COMPUTATION OF TREND MEANS

We will now consider the derivation of distributions of trend
means. The goodness-of-fit test does not dictate the manner in
which trend means are computed for the test, but a particular
procedure that has been found useful for computing trend means
will be the focus of attention in this section. That procedure is
the one associated with the trend tests employed by Price (Price
and Cooper, 1975), whose study and experimental results were
described earlier in this chapter.

The method for determining the trend means depends on the
type of trend expected. Suppose that we start with the method
to be used when the experimenter expects a linear upward trend
in the means. The first step is to get "coefficients" for the independent variable (treatment) values by subtracting the smallest

value from each of the values. For example, if the independent variable values were 20, 25, 50, 75, and 80, the corresponding coefficients would be 0, 5, 30, 55, and 60. Then we solve the following equation for a multiplier m: $m(0n_1 + 5n_2 + 30n_3 + 55n_4 + 60n_5) = GT$, where GT is the grand total of the obtained measurements for the five groups. Thus multiplier m can be computed by dividing the grand total of the obtained measurements for the five groups by the sum of the product of the sample sizes and the coefficients. When the value of m is thus obtained, it is then multiplied by the individual coefficients to obtain the *trend means*. The trend mean of the first group, then, is $0m$, the trend mean of the second is $5m$, and so on. The trend means derived by this procedure have the following two properties: (1) because of the way the multiplier m is determined, the sum of the product of the trend means and their corresponding sample sizes equals the grand total, and (2) by setting the smallest coefficient and consequently the smallest trend mean equal to 0, the trend means plotted on a graph form the steepest upward sloping line possible for nonnegative trend means. (If the data contains negative values, the data can be adjusted, before computing coefficients, by adding a constant to eliminate negative values. This, like other linear transformations of the data, does not affect the goodness-of-fit P-values.) The importance of this second property can be seen in the following example.

Example 9.4

A psychologist randomly assigns six rats to three treatment conditions, with two rats per treatment. The treatments consist of exposing an image on a screen in front of a rat for either 5, 10, or 15 seconds. According to the theory being investigated, varying the exposure time between 5 and 15 seconds should have no effect on the dependent variable. The psychologist, on the other hand, believes that the response magnitudes should be proportional to the duration of the projected image. Therefore, the trend means should be determined for the goodness-of-fit trend test by using 5, 10, and 15 as coefficients instead of using 0, 5, and 10, which provide the steepest upward trend of trend means. The experimenter multiplies the trend coefficients by $GT/[(2)(5) + (2)(10) + (2)(15)]$, where GT is the total of all six measurements, to get the trend means.

The total of the six measurements was 90, so the coefficients were multiplied by 1.5 to provide 7.5, 15, and 22.5 as the trend means. The experimental results follow:

Exposure time (seconds):	5	10	15
Trend means:	7.5	15	22.5
Measurements:	0, 7.5	15, 15	22.5, 30

Test statistic value:

$$2(3.75 - 7.5)^2 + 2(15 - 15)^2 + 2(26.25 - 22.5)^2 = 56.25$$

The measurements increase in size consistently, with no overlap, from one exposure time to the next, so, as there are $6!/2!2!2! =$ 90 data permutations, one might expect the P-value for the above results to be 1/90, but that is not the case. There are other data permutations providing test statistic values as small as 56.25. In fact, there are two data permutations providing test statistic values of 0. Consider the following data permutation:

Exposure time (seconds):	5	10	15
Trend means:	7.5	15	22.5
Measurements:	0, 15	7.5, 22.5	15, 30

Test statistic value:

$$2(7.5 - 7.5)^2 + 2(15 - 15)^2 + 2(22.5 - 22.5)^2 = 0$$

There are two 15's, and switching them provides a second permutation with a test statistic value of 0.

Note that the measurements in the data permutation that gives a test statistic value of 0, unlike the measurements for the obtained results, do *not* increase consistently, without overlap, as exposure time increases. Furthermore, the sample means do not increase as rapidly as for the obtained results. Nevertheless, if the experimenter had obtained the results represented by the data permutation with a test statistic value of 0, he would have obtained a smaller P-value than that associated with his actual results.

When the trend means are computed according to the recommendation made earlier, to provide the steepest upward trend in trend means, the P-value for the obtained results is 1/90. The trend coefficients are obtained by subtracting 5, the smallest exposure time, from 5, 10, and 15 to provide 0, 5, and 10. These trend coefficients are then multiplied by $90/[(2)(0) + 2(5) + 2(10)] = 3$, to get 0, 15, and 30 as trend means. Consider the computation of the test statistic value for the obtained results, using these trend means:

Exposure time (seconds):	5	10	15
Trend means:	0	15	30
Measurements:	0, 7.5	15, 15	22.5, 30

Test statistic value:

$$2(3.75 - 0)^2 \quad 2(15 - 15)^2 + 2(26.25 - 30)^2 = 56.25$$

The test statistic for the obtained results is the same as when trend means were 7.5, 15, and 22.5, but with the present trend means, 0, 15, and 30, no other data permutation provides as small a test statistic value as 56.25. For example, the data permutation which gave a test statistic value of 0 with the earlier set of trend means now gives a large value:

Exposure time (seconds):	5	10	15
Trend means:	0	15	30
Measurements:	0, 15	7.5, 22.5	15, 30

Test statistic value:

$$2(7.5 - 0)^2 + 2(15 - 15)^2 + 2(22.5 - 30)^2 = 225$$

The test statistic value here is large because, even though perfectly linear and upward in trend, the *measurement means* do not increase at a rapid rate as the exposure time is increased. a result of lack of a consistent upward trend in the *measurements* over exposure times.

The derivation of trend means for expected downward linear trends can be performed in an analogous manner, as also can quadratic trends. For each trend, the trend means multiplied by the corresponding sample sizes would equal the grand total of the measurements and the trend means would form the steepest trend of the specified kind for nonnegative trend means.

If the experimenter predicts a linear *downward* trend, the coefficients can be determined by subtracting each independent variable value from the largest value. For example, if the independent variable values were 10, 12, 18, and 25, the coefficients would be 15, 13, 7, and 0, respectively. These coefficients would be used to determine the trend means in the manner described earlier for a predicted upward trend.

Now consider the method for determining the trend means when the experimenter expects a U-shaped distribution of means that is symmetrical about a predetermined point of symmetry. Obtaining coefficients for the independent variable values for this kind of distribution is done by squaring the deviation of the independent variable values from the predetermined value expected to be the low point of the distribution.

Example 9.5

In Example 9.1 the predicted low point was 4 days, and so the coefficient for a training-testing interval is the square of the deviation of the training-testing interval from 4 days:

Time	30 min	1 day	3 days	5 days	7 days	10 days
Coeff	16	9	1	1	9	36

(Actually, 30 minutes is not 4 days deviation from the predicted low point of 4 days, but it is very close.) Using these coefficients, the trend means were determined by solving for the multiplier m, as shown in the second paragraph of this section. The sample sizes for the six intervals were 11, 10, 12, 8, and 10, respectively, and the grand total was 2,503.3. (Multiplying each mean in Example 9.1 by the respective sample size and summing gives a total of 2,504.4, instead of 2,503.3, because of rounding errors for the means in the example.) Consequently the following equation is solved for m: $m[(16)(11) + (9)(10) + (1)(15) + (1)(12) + (9)(8) + (36)(10)] = 2,503.3$. The value of m is then 2,503.3/725, or about 3.45. Therefore, in order of training-testing interval length, the trend means were: (3.45)(16), (3.45)(9), (3.45)(1), (3.45)(1), (3.45)(9), and (3.45)(36). That is, the trend means were: 55.20, 31.05, 3.45, 3.45, 31.05, and 124.20. If these means were plotted against their corresponding training-testing intervals, they would form a U-shaped quadratic curve with a low point of 0 at 4 days. The obtained test statistic value, then, was: $11(45.5 - 55.20)^2 + 10(40.4 - 31.05)^2 + 15(33.1 - 3.45)^2 + 12(31.9 - 3.45)^2 + 8(36.2 - 31.05)^2 + 10(43.1 - 124.20)^2 = 90,793.2$.

At this point it should be mentioned that there was no basis for predicting a particular U-shaped quadratic function for this study; the quadratic function produced by the above procedure was derived because it constituted an objective, quantitative definition of a steep, symmetrical U-shaped distribution centered at 4 days, with which the observed distribution and the distributions under all permutations of the data could be compared.

The method for determining trend means for an *inverted* U-shaped distribution about a predetermined high point is similar to that for the U-shaped distribution. The coefficients are computed as for a U-shaped distribution, and then subtracted from the largest coefficient.

Example 9.6

For the data in Example 9.5, if the experimenter had predicted an inverted U-shaped trend centered on the independent variable value of 4 days, the coefficients would be:

Time	30 min	1 day	3 days	5 days	7 days	10 days
Coeff	20	27	35	35	27	0

Each of the coefficients for a U-shaped trend test was subtracted from 36, the coefficient for 10 days. Using these new coefficients, the following equation is solved for m: $m[(20)(11) + \cdots + (0)(10)] = 2,503.3$; then each of the coefficients is multiplied by the computed value of m to get the trend means.

In addition to linear and quadratic functions, other polynomial functions, like cubic, quartic, and quintic can be used as predicted trends. When the independent variable values are uniformly spaced, published tables of polynomial coefficients can be used to generate trend means for polynomial functions.

Example 9.7

Assuming that there are no negative measurement values, if we have nine uniformly spaced independent variable values and expect a cubic trend, we could use cubic polynomial coefficients from a table of orthogonal polynomial coefficients in Winer (1971, p. 878) to generate the following goodness-of-fit coefficients:

Independent variable value	15	20	25	30	35	40	45	50	55
Polynomial coeff	−14	7	13	9	0	−9	−13	−7	14
Goodness-of-fit coeff	0	21	27	23	14	5	1	7	28

The distribution of goodness-of-fit coefficients is obtained from the polynomial coefficients by adding a constant, 14, that will make the lowest value 0. It can be seen that the derived distribution predicts an upward trend followed by a downward trend and, finally, another upward trend.

Example 9.8

If, on the other hand, it is more reasonable to expect the trend to be down-up-down, one should not use the derived coefficients but should subtract each goodness-of-fit coefficient in Example 9.7 from 28, the largest coefficient. This provides the following sequence of coefficients:

Independent variable value	15	20	25	30	35	40	45	50	55
Coefficients	28	7	1·	5	14	23	27	21	0

The new distribution of coefficients is appropriate for the down-up-down trend.

9.5   COMPUTER PROGRAM FOR GOODNESS-OF-FIT TREND TEST

Program 9.1 is satisfactory for any type of predicted trend, provided that the distribution of trend means is computed separately and fed into the computer along with the data. Program 9.1 can be tested with the following data, for which the trend means for an upward linear trend have been computed and which should give a probability value of 2/90, or about 0.0222, by systematic data permutation:

Independent variable values	100	150	200
Trend means	0	5	10
Measurements	1, 3	2, 4	9, 11

PROGRAM 9.1   Goodness-of-fit Trend
Test: Random Permutation

```
            ((INPUT))

    READ(5,---)NGRPS,(NCELL(I),I=1,NGRPS)
    READ(5,---)(TREND(I),I=1,NGRPS)
    DO 1 I=2,NGRPS
  1 NCELL(I)=NCELL(I-1)+NCELL(I)
    N=NCELL(NGRPS)
    READ(5,---)NPERM
    READ(5,---)(DATA(I),I=1,N)
    NLE=0
    DO 4 I=1,NPERM
    TEST=NN=0
    DO 3 J=1,NGRPS
    SUM=0
    N=NN+1
    NN=NCELL(J)
    DO 2 K=N,NN
    IF(I.EQ.1) GO TO 2
    ID=K+INT(RANF(0)*(NCELL(NGRPS)-K+1))
    X=DATA(K)
    DATA(K)=DATA(ID)
    DATA(ID)=X
  2 SUM=SUM+DATA(K)
  3 TEST=TEST+(NN-N+1)*(TREND(J)-SUM/(NN-N+1))**2
    IF(I.EQ.1)OBTAIN=TEST
  4 IF(TEST.LE.OBTAIN)NLE=NLE+1
    PROB=FLOAT(NLE)/NPERM

            ((OUTPUT))
```

9.6   CORRELATION TREND TEST

The goodness-of-fit trend test permits very precise predictions
of trend and can considerably increase the likelihood of detecting
a treatment effect. The procedure, however, lacks conceptual
simplicity. The goodness-of-fit test does not resemble parametric
trend analysis or, indeed, any conventional statistical test, and
the computation of the trend means complicates the procedure.
After the goodness-of-fit test had been developed, it gradually
became clear that it was possible to construct a simpler trend
test that, nevertheless, would permit precise predictions. The
new test was called the *correlation trend test* because the test
statistic is the product-moment correlation coefficient. The one-
and two-tailed test statistics are r and $|r|$, respectively, where

the correlation is between the measurements and coefficients that represent the expected trend. (Since the trend means computed according to the procedure recommended for the goodness-of-fit trend test are linearly related to the coefficients, trend means need not be used with the correlation trend test because the same P-value would be obtained as with the coefficients.) Like the goodness-of-fit trend test, the correlation trend test does not require the trend coefficients to be computed in some particular way; it is a valid randomization test procedure for any set of coefficients that is independent of the data. Nevertheless, the methods recommended for determining trend coefficients for the goodness-of-fit trend test appear to be useful for the correlation trend test, as well.

This test is both computationally and conceptually simpler than the goodness-of-fit trend test; yet it can accommodate quite specific predictions. There is no need to compute trend means; only the coefficients which define the shape of the trend are required. The logical link between this test and conventional statistical tests is clear. In the first place, this test is simply a generalization of the correlation test considered in Chapter 8. When there is only one subject assigned to each level of the independent variable, the correlation trend test is reduced to the ordinary correlation test. Secondly, the correlation trend test is related to the parametric trend test. The *numerator* of the orthogonal polynomial trend test F is equivalent under data permutation to r^2, the square of the correlation between the trend polynomial coefficients and the measurements. For the special case where the coefficients are the same as a set of polynomial coefficients (or where the two sets are linearly related) the two-tailed correlation trend test statistic is an equivalent test statistic to the numerator of the parametric trend F. An example will illustrate the employment of the correlation trend test where a linear upward trend is predicted.

Example 9.9

Suppose that we expected a linear upward trend over five levels of an independent variable, the five levels being 10, 15, 20, 25, and 30. We use the goodness-of-fit procedure to determine the following coefficients:

Treatment magnitude	10	15	20	25	30
Coefficients	0	5	10	15	20

TABLE 1   Correlation Trend Test Data

Coefficients (X)	Measurements (Y)	XY
0	23	0
0	21	0
0	20	0
5	18	90
5	25	125
5	16	80
10	12	120
10	10	100
10	13	130
15	17	255
15	10	150
15	8	120
20	12	240
20	16	320
20	4	80
$\Sigma X = 150$	$\Sigma Y = 225$	$\Sigma XY = 1810$

Suppose that there were three measurements for each independent variable value: (23, 21, 20); (18, 25, 16); (12, 10, 13); (17, 10, 8); and (12, 16, 4). The measurements would be paired with the independent variable coefficients in the way shown in Table 1. The obtained test statistic value for our one-tailed test is 1,810. Thus we could determine the proportion of the $15!/(3!)^5$ data permutations that give a value of 1,810 or larger for ΣXY, and that would be the significance for the treatment effect.

With correlation trend tests, we are concerned with the set of $N!/n_1!n_2! \cdots n_k!$ data permutations associated with random assignment to independent groups. The permutation of the data for the correlation trend test is the same as for one-way ANOVA, since the random assignment procedure is the same.

For a two-tailed test where a linear trend is expected but the direction is not predicted, the test statistic used is $|N\Sigma XY - \Sigma X\Sigma Y|$, the two-tailed test statistic for correlation. ΣX is the sum of the N coefficients for all N subjects, *not* the sum of the k coefficients for the k treatment levels. It does not matter whether the coefficients used are those for upward or for downward linear trend; the two-tailed test statistic value for any data permutation is the same for both sets of coefficients and, consequently, the P-value is the same.

Example 9.10

We have the same independent variable values as in Example 9.9, but predict a U-shaped trend, with the low value of the U at the independent variable value of 15. We compute our *coefficients*, as described in Section 9.4, for the goodness-of-fit test to get the following values:

Treatment magnitude	10	15	20	25	30
Coefficients	25	0	25	100	225

These coefficients are paired with the measurements to determine how strongly they are correlated. A strong product-moment correlation between the measurements and the coefficients implies that the magnitudes of the measurements are distributed over the independent variable levels in the same way as the coefficients: low for 15 and increasing in both directions from 15.

The two-tailed test statistic $|N\Sigma XY - \Sigma X\Sigma Y|$ would have been used in Example 9.10 if no prediction was made as to whether the trend would be U-shaped or inverted U-shaped. The two-tailed prediction, then, would be that of a large absolute value of the correlation between the coefficients and the measurements, and that prediction would be confirmed by either a strong positive or a strong negative correlation between the coefficients and the measurements. As with any predicted trend, it makes no difference for the correlation trend test which of the two sets of coefficients is used to determine the two-tailed significance.

Program 9.2 can be used to determine one- and two-tailed P-values by random data permutation for predicted trends, using the correlation trend test. It presupposes independent-groups

PROGRAM 9.2 Correlation Trend
Test: Random Permutation

```
          ((INPUT))

  TOTX=TOTY=0
  READ(5,---)NGRPS,(NCELL(I),I=1,NGRPS)
  READ(5,---)(COEFF(I),I=1,NGRPS)
  DO 1 I=1,NGRPS
1 TOTX=TOTX+(NCELL(I)*COEFF(I))
  DO 2 I=2,NGRPS
2 NCELL(I)=NCELL(I-1)+NCELL(I)
  N=NCELL(NGRPS)
  READ(5,---)NPERM
  READ(5,---)(DATA(I),I=1,N)
  DO 3 I=1,N
3 TOTY=TOTY+DATA(I)
  PRODTOT=(TOTX)*(TOTY)
  NGEONE=NGETWO=0
  DO 6 I=1,NPERM
  TESTONE=TESTTWO=NN=0
  DO 5 J=1,NGRPS
  SUM=0
  N=NN+1
  NN=NCELL(J)
  DO 4 K=N,NN
  IF(I.EQ.1) GO TO 4
  ID=K+INT(RANF(0)*(NCELL(NGRPS)-K+1))
  X=DATA(K)
  DATA(K)=DATA(ID)
  DATA(ID)=X
4 SUM=SUM+DATA(K)
5 TESTONE=TESTONE+(SUM*COEFF(J))
  TESTTWO=ABS((NN*TESTONE)-PRODTOT)
  IF(I.EQ.1)OBTONE=TESTONE
  IF(I.EQ.1)OBTTWO=TESTTWO
  IF(TESTONE.GE.OBTONE)NGEONE=NGEONE+1
6 IF(TESTTWO.GE.OBTTWO)NGETWO=NGETWO+1
  PROBONE=FLOAT(NGEONE)/NPERM
  PROBTWO=FLOAT(NGETWO)/NPERM

          ((OUTPUT))
```

---

random assignment of subjects to treatments with the treatment
sample sizes fixed in advance, the type of random assignment
used with one-way ANOVA. Whereas the program for the goodness-of-fit test required trend means to be fed into the computer,
this program requires only the coefficients to be fed in. Program
9.2 can be tested with the following data, where the coefficients
are based on a predicted downward linear trend. The one-tailed
P-value for systematic data permutation is 2/90, or about 0.0222,
and the two-tailed P-value is 4/90, or about 0.0444.

Independent variable values	30	35	40
Coefficients	28	14	0
Measurements	16, 19	14, 18	8, 9

9.7 CORRELATION TREND TEST FOR FACTORIAL DESIGNS

Tests of main effects can be carried out on data from factorial
designs when a trend is predicted for some or all of the factors.
When a factor under test does not have a predicted trend, Program 6.1 can be used in its present form, of course, to test the
effect of the factor, but to test effects of factors with predicted
trends, one must modify the program. The way in which Program 6.1 permutes the data need not be changed, but the corre-
lation trend test statistics, ΣXY and $|N\Sigma XY - \Sigma X\Sigma Y|$, which are
equivalent test statistics to r and $|r|$, would be substituted for
T_L and $\Sigma(T^2/n)$, which necessitates also a specification of trend
coefficients as part of the input to the program. We will now
describe a two-factor experimental design where one factor is a
subject variable, sex.

Example 9.11

We have a treatments-by-sex experimental design where we expect
a linear upward trend over levels of the manipulated factor for
both sexes. There are three levels of the independent variable,
and the coefficients are determined to be 0, 5, and 10. There
are six males and six females, randomly assigned to the three
treatment levels with the restriction that two males and two females take each treatment. The results are shown in Table 2.
The obtained ΣXY is $(0)(2) + (0)(4) + \cdot \cdot \cdot + (10)(12) +$

TABLE 2   Treatments-by-
Subjects Design

	Coefficients		
	0	5	10
Males	2	5	8
	4	5	9
Females	6	10	12
	7	11	14

$(10)(14) = 585$. The P-value given by systematic data permutation would be the proportion of the $(6!/2!\,2!\,2!)^2 = 8,100$ data
permutations that provide as large a value of ΣXY as 585.   No
other data permutation would provide as large a value as the obtained results, since within both sexes there is a consistent
increase in size of the scores from the lowest to the highest levels
of the treatments.   Thus the P-value would be 1/8,100, or about
0.00012.

9.8   DISPROPORTIONAL CELL FREQUENCIES

Disproportional cell frequencies in factorial designs were discussed
in Chapter 6 in connection with tests not involving a predicted
trend.   Here we will consider disproportional cell frequencies in
relation to the correlation trend test.   The basic difference between this and the earlier discussion is in the test statistic.
Earlier a special SS_B test statistic based on the deviation of a
treatment mean from the expected value of the treatment mean
over all data permutations was proposed for use with disproportional cell frequencies.   A similar modification of the two-tailed
correlation test statistic will be proposed, to provide a more
powerful test when cell frequencies are disproportional.   For a
one-tailed correlation trend test, the test statistic ΣXY is appropriate, even when cell frequencies are disproportional.
    When cell frequencies are disproportional, the two-tailed
correlation trend test statistic $|N\Sigma XY - \Sigma X\Sigma Y|$ is not as power-
ful a test statistic as when cell frequencies are proportional.
When cell frequencies are proportional, $\Sigma X\Sigma Y$ is the expected

value of $N\Sigma XY$ over all data permutations so that, if we use the expression $E(N\Sigma XY)$ to refer to the average or expected value of $N\Sigma XY$ over all data permutations, a *general* formula for the two-tailed correlation trend (and ordinary correlation) test statistic is $|N\Sigma XY - E(N\Sigma XY)|$. In computing the two-tailed test statistic for factorial designs with disproportional cell frequencies we will use the same *general* formula but we will not use $\Sigma X\Sigma Y$ as $E(N\Sigma XY)$ in the formula because then $E(N\Sigma XY)$ over all data permutations is not necessarily $\Sigma X\Sigma Y$. Instead we compute $E(N\Sigma XY)$ directly from the data.

With the correlation trend test, as with simple independent-groups or repeated-measures designs, it sometimes is useful to restrict the random assignment procedure according to the type of subject assigned, which is likely to result in disproportional cell frequencies. For religious, medical, or other reasons, it may be appropriate to restrict the random assignment to the lower levels of the independent variable, for example, whereas other subjects may be assigned to any of the levels. In the following example of restricted-alternatives random assignment, the relevance of the general two-tailed test statistic $|N\Sigma XY - E(N\Sigma XY)|$ is discussed.

Example 9.12

We have four young subjects to be assigned to levels 1 or 2 of four treatment levels, with two subjects per level, and four older subjects to be assigned to any one of the four levels, with one subject per level. There are, then, $4!/2!\,2! \times 4! = 144$ possible assignments. A linear upward trend is predicted and the results in Table 3 are obtained. We compute the means of the two age groups to get 15 for the young subjects and 9 for the old. $E(\Sigma Y)$ for the first treatment level would be the number of subjects in the top cell within that level multiplied by 15, plus the number of subjects in the bottom cell multiplied by 9, which is $(2)(15) + (1)(9) = 39$. $E(\Sigma Y)$ for the second level also is 39, computed in the same way. The $E(\Sigma Y)$ value for each of the remaining treatment levels is 9. $E(\Sigma XY)$ is the sum of the products of the coefficients and $E(\Sigma Y)$ values for the respective levels. Thus we have $(0)(39) + (10)(39) + (20)(9) + (30)(9) = 840$ for $E(\Sigma XY)$. $E(N\Sigma XY)$ is 8×840, or 6,720. The two-tailed test statistic value for the obtained results is $|N\Sigma XY - 6,720|$, which is $|7,600 - 6,720|$, or 880. There is one other data permutation that would give such a large value as 880: that where the order of the measurements over treatment levels is reversed

TABLE 3   Restricted-alternatives
Random Assignment

	Trend coefficients			
	0	10	20	30
Young	12	16		
	14	18		
Old	7	8	10	11

within each age group so that the two-tailed P-value is 2/144, or about 0.014.

It would have been perfectly valid to use $|N\Sigma XY - \Sigma X\Sigma Y|$ as the two-tailed test statistic in Example 9.12 because it could be computed over all data permutations so that, if there was no treatment effect, only chance would tend to make the value for the obtained results large relative to the other data permutations. There would have been a loss of *power*, however. Although the lowest measurements within each row separately are associated with the lowest coefficients, the highest measurements for both groups combined are associated with the lowest coefficients so that, overall, there tends to be little correlation between the coefficients and the measurements. Put differently, by restricting the random assignment of the high-scoring (young) subjects to the lower levels of treatment, the strong positive correlation between coefficients and measurements within each row separately is masked when the regular two-tailed test statistic is used. For the obtained results, $|N\Sigma XY - \Sigma X\Sigma Y|$ would be $|7,600 - 7,680|$, or only 80. No other data permutation could provide as large a value of $N\Sigma XY$ as 7,600, and so every data permutation would have a value of $|N\Sigma XY - \Sigma X\Sigma Y|$ as large as the obtained results. Consequently the two-tailed significance value would have been 1.

It can be seen that when there are disproportional cell frequencies the general test statistic $|N\Sigma XY - E(N\Sigma XY)|$ can be much more powerful than $|N\Sigma XY - \Sigma X\Sigma Y|$. For a randomization test, the general test statistic is equivalent to the simpler test statistic $|\Sigma XY - E(\Sigma XY)|$, the absolute difference between ΣXY for a data permutation and the expected or average value of ΣXY over all data permutations. This in turn is equivalent to $|r - E(r)|$, the absolute difference between the product-moment

correlation coefficient for a data permutation and the mean correlation coefficient over all data permutations. In other words,
for determining significance by the randomization test procedure,
the general two-tailed test statistic is equivalent to the absolute
difference between a value of r and the value that would be
expected by "chance." When cell frequencies are proportional,
$E(r)$ becomes 0, making the special test statistic then equivalent
to $|r|$.

9.9 DATA ADJUSTMENT FOR DISPROPORTIONAL CELL FREQUENCY DESIGNS

In Section 6.14 it was shown that an adjustment of data could
serve the same function as modifying the two-tailed test statistic
in making a randomization test more sensitive to treatment effects
when cell frequencies are disproportional. The same data adjust-
ment can be used before computing $|N\Sigma XY - \Sigma X\Sigma Y|$ for a cor-
relation trend test to achieve the same P-value as that of the
special test statistic $|N\Sigma XY - E(\Sigma XY)|$. Program 9.1, without
modification, can be applied to the adjusted data.

Example 9.13

To provide a way of determining significance for a correlation
trend test applied to factorial designs, we modify Program 6.1
by adding trend coefficients as inputs and changing the test
statistics from T_L and $\Sigma(T^2/n)$ to ΣXY and $|N\Sigma XY - \Sigma X\Sigma Y|$. We
use these test statistics whether the designs have proportional or
disproportional cell frequencies, but when the designs have disproportional cell frequencies the data are adjusted before being
subjected to the statistical test. The adjustment carried out on
the data in Table 3 (in Example 9.12) would be to express
measurements for young subjects as $Y - \bar{Y}$ and measurements for
old subjects as $O - \bar{O}$, where $\bar{Y}$ and $\bar{O}$ are the means for the two
age groups. Table 4 shows the results of the study in terms
of adjusted data. First, let us examine the relationship between
one-tailed P-values for r for the unadjusted measurements (Table
3) and the adjusted measurements (Table 4). The adjusted
measurements are linearly related to the unadjusted measurements
for the young subjects, and the adjusted measurements are linearly related to the unadjusted measurements for the old subjects.
One should not, however, expect r to be the same for the adjusted and unadjusted data because the linear relationship between

TABLE 4   Adjusted Data

	Trend Coefficients			
	0	10	20	30
Young	−3	+1		
	−1	+3		
Old	−2	−1	+1	+2

adjusted and unadjusted measurements is different for the old and young subjects. Nevertheless, it can be shown that r for adjusted data is an equivalent test statistic to r for unadjusted data, providing the same P-value.

For either adjusted or unadjusted data, ΣXY is an equivalent test statistic to r, so to show the equivalence of r's for adjusted and unadjusted data, it is sufficient to demonstrate that ΣXY for unadjusted data (ΣXY_{un}) and ΣXY for adjusted data (ΣXY_{adj}) are equivalent test statistics. The demonstration of this equivalence will be accomplished by showing that $\Sigma XY_{un} - \Sigma XY_{adj}$ is constant over all data permutations.

The following argument depends on an interesting fact about the generation of data permutations: all n! permutations of a sequence of n things can be produced by repeatedly exchanging the sequential position of a pair of things, which implies that all divisions of measurements among any number of treatment groups can be produced, one at a time, by exchanging the group membership of two of the measurements. For example, a new data permutation could be derived from the data permutation shown in Table 4 switching −3 and +1 in the young group, and then a data permutation could be derived from that one by moving +1 again, this time exchanging its position with that of +3. This process of switching pairs of numbers within the young group and within the old group could be continued until all 144 data permutations were represented. If we listed the data permutations in the order in which they were produced, each data permutation would be the same as the one preceding it except for the location of two measurements.

The amount of change in ΣXY resulting from exchanging two measurements, a and b, in the same row, between columns for a higher coefficient (HC) and a lower coefficient (LC) is $|HC -$

$LC| \times |a - b|$, and ΣXY is reduced if a and b are negatively correlated with the coefficients after switching, and ΣXY is increased if a and b are positively correlated with the coefficients after switching. Now, the difference, $a - b$, is the same for two measurements, whether a and b refer to unadjusted measurements or the corresponding adjusted measurements. Suppose two series of 144 data permutations have been produced by successive switching of two measurements for the unadjusted data, starting with the obtained results, and simultaneously switching the corresponding measurements for the adjusted data, starting with the obtained adjusted data. Then since the difference, $a - b$, is the same for two measurements, whether the measurements are adjusted or unadjusted, ΣXY_{adj} and ΣXY_{un} must increase or decrease by the same amount between the first data permutation and the second, increase or decrease equally between the second and third data permutations, and so on for all successive data permutations. Thus, $\Sigma XY_{un} - \Sigma XY_{adj}$ is constant over all data permutations, so the two one-tailed test statistics are equivalent, giving the same P-values.

Let us turn to consideration of the two-tailed test statistic $|N\Sigma XY - \Sigma X\Sigma Y|$, the test statistic whose apparent inappropriateness for factorial designs with disproportional cell frequencies led to the modified two-tailed test statistic, $|N\Sigma XY - E(N\Sigma XY)|$. We will show in six steps that even though $|r|$ for unadjusted data may be different from $|r|$ for adjusted data, the test statistics $|N\Sigma XY_{adj} - \Sigma X\Sigma Y_{adj}|$ and $|N\Sigma XY_{un} - E(N\Sigma XY_{un})|$ are numerically equal. We start with the relationship between the one-tailed test statistics

$$\Sigma XY_{un} - \Sigma XY_{adj} = c \tag{9.1}$$

where c is a constant over all data permutations. From (9.1), we derive

$$N\Sigma XY_{adj} = N\Sigma XY_{un} - Nc \tag{9.2}$$

From (9.2), we get the following equation of expectations:

$$E(N\Sigma XY_{adj}) = E(N\Sigma XY_{un}) - Nc \tag{9.3}$$

Because the sum of adjusted measurements must equal 0, $E(N\Sigma XY_{adj}) = 0$, so

$$E(N\Sigma XY_{un}) = Nc \tag{9.4}$$

Substituting $E(N\Sigma XY_{un})$ for Nc in (9.2) we get

$$N\Sigma XY_{adj} = N\Sigma XY_{un} - E(N\Sigma XY_{un}) \tag{9.5}$$

Since the sum of the adjusted measurements is 0, $\Sigma X\Sigma Y_{adj} = 0$, so $\Sigma X\Sigma Y_{adj}$ can be subtracted from the left side of (9.5), leading to the following equation, which is the expression of the numerical identity of the two-tailed test statistics for adjusted and unadjusted data:

$$\left| N\Sigma XY_{adj} - \Sigma X\Sigma Y_{adj} \right| = \left| N\Sigma XY_{un} - E(N\Sigma XY_{un}) \right| \tag{9.6}$$

Thus, applying Program 9.2 to the adjusted data will provide the same two-tailed P-value as modifying the program and applying it to the raw data. Both test statistics are equivalent to $|r - E(r)|$ for the respective data types, and as $E(r_{adj})$ equals 0, both test statistics are equivalent to $|r_{adj}|$.

9.10   REPEATED-MEASURES TREND TESTS

Repeated-measures experiments can be regarded as randomized block experiments where each block is a separate subject. The discussion of correlation trend tests for factorial designs in preceding sections of this chapter, such as restricted-alternatives random assignment and the need for special computational procedures when cell frequencies are disproportional, thus applies to repeated-measures trend tests as well as to trend tests for other factorial designs.

9.11   CORRELATION TREND TEST AND SIMPLE CORRELATION

When there is only one subject for each of k levels of a quantitative independent variable, the test statistics are equivalent to the correlation between the k coefficients and the k measurements. Coefficients for linear trend are linearly related to the independent variable values, and so the P-values are the same for the correlation trend test for linear trend with the use of the linear trend coefficients as with the independent variable magnitudes

used as the coefficients. Thus, simple correlation described in
Chapter 8 gives the same P-value as if the data were tested by a
correlation trend test for linear trend. Simple correlation, then,
is a special case of the correlation trend test: the case where
there is only one subject per level of treatment and a linear
trend is predicted.

There is no reason, however, why only a linear trend can
be tested when there is just one subject per treatment level.
The coefficients that are determined are independent of the number of subjects per treatment level, and so they are the same for,
say, a U-shaped trend with a low point at the independent variable value of 15, whether there are 20 subjects for each level of
the independent variable or only one subject. Thus, the use of
the correlation trend test with data where there is only one subject per treatment level is effectively an extension of product-
moment correlation to situations where any type of trend can be
predicted. The same degree of specificity of prediction can be
accommodated as if there were several subjects at each treatment
level. With one subject per treatment, therefore, the coefficients
can be determined for the predicted trends and used as the X
values to be paired with measurements in Program 8.1 or 8.2.

9.12 ORDERED LEVELS OF TREATMENTS

The correlation trend test can be applied even when the levels
of the independent variable are not quantitative, provided that
they can be ordered with respect to the expected trend in the
data.

Example 9.14

For short periods of time, people apparently are able to ignore
or block out distractions that otherwise would impair their performance on a skilled task. One would expect, however, that
over long periods of time the effort to carry out such blocking
would be too great to sustain and that as a consequence, the
performance would suffer. An experiment is designed to determine whether the noise of other typewriters and movement in the
periphery of the field of vision, of a level comparable to that in
a typical office, would adversely affect the quality of typing if
the noise and the visual distraction were continuous over several
hours. The quality of typing is operationally defined as the
typing speed in words per minute, adjusted for errors. Three

levels of the independent variable are used: N (no visual or auditory distraction), A (auditory distraction), and A + V (auditory and visual distraction). The null hypothesis is: the typing speed of each typist is independent of the experimental conditions. We expect N to provide the highest measurements, A + V to provide the lowest, and A to provide measurements of intermediate size; consequently the coefficients for the correlation trend test are ranks 1, 2, and 3, assigned in the following way:

Treatment level	A + V	A	N
Coefficients	1	2	3

Six typists with similar experience are selected for the experiment, and two are randomly assigned to each treatment level, there being 6!/2!2!2! = 90 possible assignments. The following typing speeds were obtained: A + V: 53, 52; A: 56, 55; and N: 59, 60. Program 9.2 can be used to determine the P-value. To compute the one-tailed test statistic ΣXY for the obtained data, the computer pairs the measurements with the coefficients as in Table 1 for the data in Example 9.9. For the obtained data in the present example, $\Sigma X = 12$, $\Sigma Y = 335$, and $\Sigma XY = 684$. ΣX and ΣY are constant over all 90 data permutations, but the pairing of the X and Y values, and therefore ΣXY, varies. No other data permutation gives a value of ΣXY as large as 684, the obtained value, and so the P-value for *systematic* data permutation would be 1/90, or about 0.011. Program 9.2, however, uses random data permutation to determine the significance of the results and may not give a P-value that is exactly 1/90. It should be kept in mind that the random data permutation procedure is valid no matter how small the sample size or how few distinctive data permutations are possible. Thus, to increase the power of the test for the present example, increasing the number of random data permutations (NPERM) to several thousand is appropriate although the systematic permutation procedure would employ only 90 permutations.

If a systematic data permutation program for the correlation trend test is desired, it can be produced by combining the part of Program 4.1 that systematically permutes the data for one-way ANOVA with the part of Program 8.1 that computes the test statistics for product-moment correlation. Program 9.2 is composed of a similar combination, using the production of test

statistics for product-moment correlation in conjunction with the
random permuting component of Program 4.3, which is for one-
way ANOVA with significance determined by random permutation.

9.13   RANKED AND DICHOTOMOUS DATA

If the data are in the form of ranks, the correlation trend test
can be applied to the ranks as if they were raw measurements.

Example 9.15

As in Example 9.14 two of six typists are randomly assigned to
each of the three conditions specified in that example: N (no
visual or auditory distraction), A (auditory distraction), and
A + V (auditory and visual distraction). The null hypothesis
is: cheerfulness in independent of experimental conditions. An
independent abserver ranks the typists with respect to cheerfulness after the experiment, assigning a rank of 1 to the least
cheerful and 6 to the most cheerful. N is expected to produce
the highest ranks, A + V to produce the lowest, and A to produce the intermediate ranks. The coefficients for the correlation
trend test thus are the same as in Example 9.14

Treatment level	A + V	A	N
Coefficients	1	2	3

The obtained cheerfulness ranks were: A + V: 1, 2; A: 3, 4;
and N: 5, 6. ΣXY for the obtained results is $(1)(1) + (1)(2) + (2)(3) + (2)(4) + (3)(5) + (3)(6) = 50$. Of the 90 data permutations, only one gives a value of ΣXY as large as 50, and so the
P-value is 1/90, or about 0.011.

Although in Example 9.15 the treatment coefficients as well
as the data are ranks, the same procedure could be followed if
only the data were ranked; for a quantitative independent variable, for example, the coefficients could be 0, 5, and 10, although the data consisted of ranks. Furthermore, tied ranks
cause no problem with regard to validity when significance is
determined by data permutation. It would be relatively easy to
construct tables for a rank-order correlation trend test where
both the data and the treatment coefficients were ranks, with no
tied ranks within either the treatment coefficients or the data,

and with equal sample sizes. With the availability of Program 9.2, however, that is unnecessary. Program 9.2 has great versatility in determining significance for ranked data: the treatment coefficients need not be ranks, tied ranks for either coefficients or data are acceptable, and the sample sizes can be equal or unequal.

Example 9.16

If the dependent variable is dichotomous, like lived-died or correct-incorrect, the numerical values 0 and 1 can be assigned to designate the categorical response and those numerical values can then be used as the quantitative data for trend analysis. Suppose that we have three levels of dosage (5, 10, and 15 units) of a pesticide and want to test its lethality on six rats. Two rats are randomly assigned to each level, providing $6!/2!2!2! = 90$ possible assignments. A linear upward trend in the proportion of rats killed by the dosages is predicted, and so the coefficients 0, 5, and 10 are associated with the treatments. The following results, where 0 indicates that a rat lived and 1 indicates that it died, were obtained: 5-unit dosage: 0, 0; 10-unit dosage: 0, 1; and 15-unit dosage: 1, 1. The obtained ΣXY, then, is $(0)(0) + (0)(0) + (5)(0) + (5)(1) + (10)(1) + (10)(1) = 25$. There is no data permutation with a larger value of ΣXY than 25, but there are nine data permutations with that value, and so the P-value is 9/90, or 0.10. None of the 5-unit dosage rats died, half of the 10-unit dosage rats died, and all of the 15-unit dosage rats died, so the results could not have been more consistent with the prediction; yet the P-value associated with the results is not at all small. Although the correlation trend test can be used with dichotomous data, the sensitivity of the test is considerably less than with continuous data.

9.14   REPORTING RESULTS OF STATISTICAL TESTS

Both the goodness-of-fit and the correlation trend tests are fairly complex as well as new and, consequently, will require more explanation in a research report than a familiar test for which significance has been determined by means of a randomization test. Except for the restricted-alternatives assignment designs, the random assignment is conventional and requires no special explanation for the reader. It is the multistage computation of the test statistic for the two trend tests that complicates their description in reporting experimental results. The goodness-of-fit test involves three stages of computation: trend coefficients, trend means, and then the test statistic. The correlation trend test involves two stages of computation: trend coefficients and the test statistic.

The correlation trend test will be easier than the goodness-of-fit trend test for the experimenter to explain. This is a consequence of the conceptual, rather than the computational, simplicity of the correlation trend test. It is easier for the experimenter to understand, and tests that are easy for the experimenter to understand are likely to be easy for the reader of his research report to understand and, therefore, will require less explanation.

The experimenter should explain each of the stages of computation. For the correlation trend test this is fairly simple. The computation of the trend coefficients used in the experiment will be easy to demonstrate and it will be easy to explain their purpose. The trend of these coefficients over the independent variable levels represents the predicted trend of the measurements over the independent variable levels. Product-moment correlation is used to determine the correlation between the coefficients and the measurements, and the higher the correlation the more similar the predicted trend (represented by the coefficients) and the actual trend (represented by the data). The significance given by the randomization test is the proportion of the data permutations providing such a large correlation coefficient as the obtained value.

When restricted-alternatives random assignment is used, the assignment procedure should be described in the research report, and it should be pointed out that the randomization test permutes the data in accordance with the type of random assignment performed. The one-tailed test statistic ΣXY is equivalent to the product-moment correlation coefficient, and the P-value associated with it is the proportion of the data permutations providing as large a correlation coefficient (taking the sign of the correlation into consideration) if a positive correlation is predicted, or as small a correlation coefficient if a negative correlation is predicted. The P-value for the two-tailed test is the proportion of the data permutations providing a correlation between the trend coefficients and the data that deviates from the average correlation over all data permutations by as much as the obtained correlation coefficient deviates. (For adjusted data, this proportion is the same as the proportion of data permutations with a value of $|r_{adj}|$ as large as the obtained value.) It is unnecessary to explain the actual test statistics that were used in the computer program and how the expected or average value of any of the test statistics is determined. These are technical details that do not affect the interpretation of the test results.

REFERENCES

Edgington, E. S. (1975). Randomization tests for predicted trends. *Can. Psychol. Rev. 16*, 49–53.

Huppert, F. A., and Deutsch, J. A. (1969). Improvement in memory with time. *Q. J. Exp. Psychol. 21*, 267–271.

Price, M. T. C., and Cooper, R. M. (1975). U-shaped functions in a shock-escape task. *J. Comp. Physiol. Psychol. 89*, 600–606.

Winer, B. J. (1971). *Statistical Principles in Experimental Design* (2d ed.). McGraw-Hill, New York.

10

Single-subject Randomization Tests

10.1 THE IMPORTANCE OF SINGLE-SUBJECT EXPERIMENTS

There are many reasons why single-subject experiments are a worthwhile undertaking. There are practical as well as theoretical advantages in conducting single-subject rather than multiple-subject experiments. In documenting a number of important single-subject studies carried out by early psychologists, Dukes (1965) discussed some of those advantages.

The rareness of certain types of experimental subjects may necessitate conducting an experiment whenever such a subject is available. With luck, an experimenter doing research on pain may recruit a person with congenital absence of the sense of pain but not be able to locate enough such persons at one time for a multiple-subject experiment. Cases of multiple personality also are rare and the psychiatrist who has such a patient probably would be interested in carrying out an intensive study on the patient.

In addition to the rareness of certain kinds of individuals, there are other compelling reasons for the intensive study of a single subject. The difficulty and expense of studying several subjects over a long period of time are good reasons for the intensive study of individual subjects in longitudinal studies. For example, a single ape can be raised with a family and be carefully studied, but such an undertaking involving several apes simultaneously is not feasible.

Their relevance to theories that concern individual organisms is by no means a trivial advantage in using single-subject experiments. Although social, economic, and political theories frequently concern the aggregate behavior of groups of people, they must take into consideration factors that influence the development of opinions, attitudes, and beliefs within individual persons. Educational, psychological, and biological theories more openly show their primary concern with individual behavior, and those theories explain phenomena in terms of processes within individuals. Experiments in various fields, therefore, are intended to further our understanding of intraorganismic processes. Typically, a measurement is taken from each subject rather than there being a single measurement from an entire group of subjects. In multiple-subject experiments, however, treatment effects on individuals sometimes are concealed by the averaging of data over a heterogeneous group of subjects. For example, if A is more effective than B for some subjects and is less effective than B for about the same number of subjects, a difference between means may be so small as to obscure treatment effects. When virtually all subjects are affected in the same manner, averaging does not have that depressing effect on statistical significance but it still reflects only a general trend. The problem with averaging over subjects is not so much that of overlooking the existence of subjects with atypical responses, although that is a problem, but inability to *identify* the individuals who respond in one way and those who respond differently in order to explore the basis of the variation in type of response. To attain an understanding of the internal processes of individuals, some experimenters prefer a series of single-subject experiments to a single multiple-subject experiment, when feasible.

Research performed on one person at a time is important for reasons aside from the scientific importance of the findings. The importance of issues relating to human rights and individual freedom throughout the world today indicates the need for precise determination of the needs and abilities of individual persons, because it is only in relation to a person's needs and abilities that it makes sense to refer to that person's rights or freedom. The demand to be treated as a person and not just a number is more than a protest against being treated in an impersonal manner; it is a demand that that person's individual characteristics that distinguish him from other people be taken into consideration. Educational techniques have been custom-made to suit the needs of students with special handicaps. Tests of various kinds are useful in developing an instructional setting appropriate for an individual student. In the field of medicine, a general practitioner who has known a patient for many years will use his knowledge of the patient to adjust his treatment accordingly. Knowing that what works for one patient may not work for another, a physician will try out one medication and another to find out what is best for the particular patient being treated. Each person uses the experience of others as a guide but spends his lifetime gradually discovering what he personally likes and what is best for him in education, medicine, occupations, recreation, and other aspects of life. In light of the importance of taking individual differences into consideration in all aspects of life, we should not rely solely on unsystematic investigation to determine the idiosyncratic "laws" that regulate the life of a particular person. Loosely controlled empirical research and crude trial-and-error approaches ("take this new drug for a week and let me know if it helps") are not enough; even though some kinds of information are attainable only by other means, single-subject experiments should supplement or replace many of the procedures currently in use for intensive investigation of individuals.

10.2   RARENESS OF PUBLISHED SINGLE-SUBJECT EXPERIMENTS

Although single-subject studies sometimes are reported in journals, they usually concern observations in natural settings, where there is no experimental manipulation of variables. Reports of *experimental* studies of individual subjects are rare. Editors expect experimental results to be evaluated by statistical tests, and reference books provide no guide for the application of a statistical test to a number of measurements from one subject— except, of course, when there are several subjects in an experiment with repeated measures on each subject. Inasmuch as single-subject experiments usually are not even mentioned in statistics books, the reason for their neglect can only be inferred. It appears likely, however, that single-subject statistical tests are uncommon mainly because of ignorance of the possibility of a valid test of that kind. After all, it is recognized generally that a statistical inference about a population depends on an estimate of the population variability, and such an estimate cannot be made on the basis of data from only one subject. In other words, it appears that the concern of statisticians with the random sampling model for drawing inferences about populations is the main reason for the lack of consideration given to single-subject experiments in statistics and experimental design books, and this lack of consideration in turn has limited the number of single-subject experiments that are published and, probably, the number that are conducted.

In Chapter 1, it was mentioned that the use of ANOVA with data from only one subject has been under criticism because of its questionable validity for such application. A number of papers and symposia for practitioners of behavior modification have dealt with the application of ANOVA to single-subject data, some presentations favoring and others opposing its application (Gentile et al. 1972; Hartmann, 1974; Keselman, 1974). The assumption questioned most strongly has been the assumption of the independence of observations. When a number of observations are taken from the same subject, observations that are close to each other in time are likely to be more similar than ones that are far apart. This serial correlation in a temporal series of observations from an individual subject violates the assumption of independence underlying the F table, in the view of some psychologists. Others, however, contend that their particular application of ANOVA has been rendered valid through taking the serial correlation into consideration in the test. So far, the tests that have received the most intensive examination of their underlying assumptions, in regard to single-subject experimental data, have been one-way ANOVA and various types of time-series analysis. It is noteworthy that, in the extensive discussion of independence of observations, the basic assumption of random sampling has been ignored completely. In the following analysis of the random sampling model it will be shown that the use of published significance tables based on the random sampling model (i.e., tables for ANOVA, time-series analysis, or any other parametric statistical technique) is invalid for single-subject experiments.

10.3   RANDOM SAMPLING MODEL

The application of statistical tests to single-subject experimental data cannot be justified on the basis of the random sampling model, which is the model underlying the probability tables for parametric tests. Random selection of a single subject from a population of subjects does not permit an estimate of the population variance, and so this type of random sampling obviously provides no basis for application of a parametric test. Another type of random sampling, that of random selection of *treatment times* from a population of such times, could be performed, but not in the manner required by the random sampling model. For example, the experimenter could designate a large number of potential treatment times, and then randomly select so many of them for one treatment, so many for another, and so on. The number of potential treatment times in the population would have to be large relative to the number selected for the various treatments, because the random sampling model involves an infinite population, not one that is almost exhausted by the sampling. On the other hand, taking small random samples from a very large population of treatment times, although inconvenient because of the likelihood of the sample treatment times being spread over a long interval of time, nevertheless is possible. However, there is an assumption of independence associated with the random sampling model that would make such samples inappropriate for testing hypotheses.

The random sampling model underlying parametric significance tables assumes that the measurement associated with a randomly selected element is independent of other elements in the sample. Thus, in reference to randomly selected treatment times for a subject, the independence assumption implies that a subject's measurement associated with a treatment at a particular time t_k is independent of the number of treatments given prior to t_k. Whether t_k is the earliest time in a sample of treatment times or the latest, then, is assumed to have no effect on a subject's response at time t_k. The inappropriateness of this assumption, because of factors like fatigue and practice, indicates that random sampling of treatment times cannot provide a basis for the valid use of parametric statistical tables with single-subject experimental data.

Thus neither random sampling of a population of subjects nor of a population of treatment times would justify the determination of significance by parametric probability tables for single-subject experimental data. Apparently there is no random sampling that can be performed that would make significance determination by reference to parametric significance tables valid for single-subject experiments.

10.4   RANDOM ASSIGNMENT MODEL

On the other hand, statistical tests whose significance is based on random assignment can be applied to single-subject experimental data validly when there has been random assignment of treatment times to treatments. A procedure of random assignment that sometimes can be employed in a single-subject experiment is one that is analogous to the random assignment of subjects to treatments in a multiple-subject experiment. In the typical multiple-subject independent-groups experiment, the assignment procedure is: the number of subjects per treatment is fixed, and within that restriction there is random assignment of subjects to treatments. The single-subject analogue is as follows: the number of treatment times for each treatment is fixed, and within that restriction there is random assignment of treatment times to treatments. Given such random assignment in a single-subject experiment, a randomization test for determining significance can be employed to test the following H_0: for each of the treatment times, the response is independent of the treatment given at that time.

The randomization tests that have been described for determining significance for ANOVA and other techniques based on random assignment of subjects to treatments therefore can be applied to single-subject experiments employing analogous assignments of treatment times to treatments. We will now consider various applications of this kind. The examples will show how the random assignment is performed and how the statistical test is to be conducted and the results interpreted.

10.5 FISHER'S CLASSICAL SINGLE-SUBJECT EXPERIMENT

In 1935 Fisher gave what is thought to be the earliest discussion of a statistical test for a single-subject randomized experiment (Fisher, 1951, Chapter 2). In order to examine its rationale, we will review Fisher's hypothetical lady-tasting-tea experiment, which has become a classic. The experiment was a test to determine whether a lady could tell by tasting a cup of tea whether tea was poured on top of milk or milk was poured on top of tea before being mixed in the cup. Eight cups were prepared, four in each manner. The cups were presented to the lady in a random order with her knowledge that the order of presentation was random, that four cups had been mixed in each way and that she was to decide which cups had milk put in first and which had tea put in first. In this hypothetical experiment, the lady correctly identified all eight cups. There are $8!/4!4! = 70$ ways eight cups can be divided into two groups

of four cups each, with one group being designated as the
"milk first" group and the other as the "tea first" group.
Fisher thus gave a P-value of 1/70 to the results because that
is the probability of correctly identifying all eight cups by
chance, that is, by randomly dividing the eight cups into "milk
first" and "tea first" groups.

In discussing the lady-tasting-tea experiment, Fisher put
emphasis on the use of randomization to control for various
sources of confounding and on aspects of the experimental
design that could be modified to increase the sensitivity of the
test to treatment effects but did not stress the procedure used
for determining statistical significance. The random order of
presentation of the cups to the lady provides the basis for a
randomization test where the data ("milk first" and "tea first"
judgments) are permuted over the treatments (cups of tea prepared by putting in milk first and then adding tea or by pour-
ing the ingredients into the cup in the opposite order), but
the procedure was not discussed from the perspective of randomization tests. As Fisher indicated, the significance was determined by a straightforward application of the formula for the
number of combinations of n things taken r at a time. That
formula was known long before Fisher's time, so Fisher was not
demonstrating a new procedure for determining statistical
significance.

10.6   THE SINGLE-SUBJECT RANDOMIZATION
TEST: AN EARLY MODEL

The "Fisher randomization test," which was applied to Darwin's
data on descendents of cross-fertilized and self-fertilized plants
(See Section 1.13) was discussed in the chapter following the
lady-tasting-tea experiment (Fisher, 1951, Chapter 3). That
fact raises the question of why Fisher did not discuss this
experiment from the randomization test perspective, as, for
example, by using a quantitative response and explicitly deriving a reference set of test statistic values based on the data.
A consideration that may help explain Fisher's failure to discuss
the lady-tasting-tea significance-determining procedure as
another instance of a randomization test is that the "Fisher randomization test" applied to Darwin's data, as pointed out in
Section 1.13, required random sampling, and random sampling
in addition to random assignment could be difficult to introduce
in a plausible manner into the lady-tasting-tea experiment.

(It should be kept in mind that it was Pitman (1937, p. 129), not Fisher, who demonstrated the applicability of randomization tests in the absence of random sampling.) Whatever the cause may be, Fisher did not present the lady-tasting-tea experiment as a new procedure for deriving a reference set of test statistics based on data permutation.

It was more than 30 years after publication of Fisher's experiment before what seems to be the first single-subject randomization test was published (Edgington, 1967). The general nature of this test, which is an independent t test applied to single-subject data with the P-value based on data permutation, and the experimental design for which it is appropriate, are illustrated in Example 10.1.

Example 10.1

In an experiment we predict that our subject will provide larger measurements under treatment A than under treatment B. In planning the experiment, we decide to administer each treatment six times. We confer with the subject to determine when the subject could be available for the experimental sessions. An afternoon is selected during which the treatments can be given. Then we select 12 times from that afternoon for treatment administration, ensuring that the treatments are far enough apart to serve our purpose. The 12 times are then assigned randomly to the treatments. The null hypothesis is: the subject's response at any treatment time is independent of the assignment of treatment times to treatments. We obtain the following data, listed in the order in which the treatments were given:

A	B	B	A	B	A	A	B	A	B	A	B
17	16	15	18	14	17	18	14	20	13	19	15

The one-tailed test statistic is T_A, the total of the A measurements, and the obtained test statistic value is 109. To perform our statistical test, we determine the proportion of the $12!/6!6! = 924$ data permutations that have a value of T_A as large as 109, using Program 4.4, the program for the independent t test which uses systematic data permutation.

We have considered a hypothetical application of the single-subject randomization test in its simplest form. Let us now examine some applications of it that have been published.

Example 10.2

A study was carried out to determine the effect of artificial
food colors on aversive (undesirable) behavior in children, such
as whining, running away, or breaking and throwing things.
(Weiss et al., 1980). The children in the study had not been
diagnosed as hyperactive, but the researchers expected artificial
food additives to have a similar effect on their behavior to that
which additives have been observed by some investigators to
have on hyperactive children.

The study consisted of 22 single-subject experiments, each
carried out on a child between 2 1/2 and 7 years of age. On
each of 77 days a child consumed a bottle of soft drink containing either natural or artificial food coloring. The authors stated
that "the two drinks were indistinguishable by sight, smell,
taste, or stain color," and neither the children nor their parents,
whose record of the child's behavior was the principal source of
data, knew which drink was given on any particular day. Eight
days were randomly selected for giving the artificially colored
drink, and the other soft drink was given on the rest of the
days. (Using a large number of control days to compensate for
a small number of experimental days, as in this study, can be a
useful tactic for providing a relatively sensitive test of treatment effect in a single-subject experiment when frequent administration of the experimental treatment is undesirable.)

Ten different measures served as dependent variables, many
of them consisting of the frequency of occurrence of a certain
type of aversive behavior. Program 4.5, using 10,000 data permutations, was applied to determine significance for each child
on each dependent variable. "Twenty of the children displayed
no convincing evidence of sensitivity to the color challenge"
(p. 1488). A 3-year-old boy had P-values of 0.01 and 0.02 for
two types of aversive behavior, which is not very impressive,
considering the multiplicity of testing: 10 dependent variables
for each of 22 children. A 34-month-old girl, however, provided very persuasive results, even when the number of children tested is taken into account. The P-value for three of the
ten dependent variables for that child was 0.0001, and P-values
for five others were 0.0003, 0.0004, 0.0006, 0.001, and 0.03.
The authors stressed the need for more stringent testing of
additives, because the food additive used in the experiment was
"about 50 times less than the maximum allowable intakes (ADI's)
recommended by the Food and Drug Administration" (p. 1488).
The single-subject experimental results of an exceptionally

sensitive child showed an effect of a small dose of food coloring that might have been missed in other experimental designs.

There were 22 randomized single-subject experiments and separate analyses of data from each of the experiments in the study described in Example 10.2. In Example 10.3, which follows, only one subject participated in the study as the objective was to determine the appropriateness of a certain treatment for that particular subject.

Example 10.3

A single-subject randomized experiment was conducted on a patient who had been taking metronidazole for three years following an operation for ulcerative colitis (McLeod, Taylor, Cohen, and Cullen, 1986). The physician and her patient agreed on the desirability of performing an experiment to determine whether the medication was relieving unpleasant symptoms, like nausea and abdominal pain, because if metronidazole was not effective, both the cost of the drug and risk of cancer associated with its long-term use dictated discontinuing it.

The experiment was of twenty weeks duration. The twenty weeks were divided into ten blocks of two-week intervals, and five of the intervals were randomly selected as the times when the patient received metronidazole daily, and the remaining five two-week intervals were assigned to "an identical placebo capsule." The primary source of data was a diary in which the patient was asked to report the presence or absence, or the magnitude, of each of seven symptoms, which included nausea, abdominal pain, abdominal gas, and watery stool. Because of the possibility that the effects of the drug would carry over to days after it was withdrawn, only data from the second week of each two-week session were subjected to a statistical test. Daily observations were combined to provide a single weekly measurement for each different symptom. There were, therefore, 10 measurements for each dependent variable (symptom), and, since there were only $10!/5!5! = 252$ data permutations, the experimenters carried out a systematic randomization test, using the data permuting procedure in Program 4.4.

Six of the seven symptoms showed a reduction under metronidazole significant at the 0.05 level: two P-values were 0.024 and four were 0.004. As the results confirmed the effectiveness of the drug, the patient was advised to continue taking it.

10.7 THE CONCEPT OF GUESSING

An element of the lady-testing-tea example (Fisher, 1951, Chapter 2) that complicates data analysis is the belief, which Fisher seems to have held, that guessing on the part of the subject introduced a chance element into the experiment in addition to that resulting from random presentation of the cups of tea. He referred to a situation in which all cups made with milk poured first had sugar added, but none with tea poured in first had sugar added, noting that this could lead to an easy division into two groups on the basis of the flavor of sugar alone, and remarked:

> These groups might either be classified all right or all wrong, but in such a case the frequency of the critical event in which all cups are classified correctly would not be 1 in 70, but 35 in 70 trials, and the test of significance would be wholly vitiated. (p. 18).

There is no reference to a random element being employed in determining that all cups with milk poured first would have sugar added and all cups with tea poured first would not have sugar added. The probability of 35/70, therefore, must be based on the assumption that after distinguishing four cups made with sugar from four made without sugar there is a probability of 1/2 of correctly designating the groups as "milk first" or "tea first." Thus, guessing is assumed to have a chance basis. No matter what Fisher *meant*, what he *wrote* represents a point of view that unjustifiably treats responses under the null hypothesis as being randomly selected. As will be shown below, the everyday notion of guessing as being like mentally tossing a coin must be abandoned before a randomization test can properly be applied to data from tests of perceptual ability.

Careful consideration of the random assignment and the corresponding data permutation procedure, along with H_0, is especially important in connection with experiments where "guessing" is the hypothesis to be rejected. There is a strong temptation to regard the chance element as residing in the subject rather than in the random procedure introduced by the experimenter. Bear in mind, however, that in the lady-tasting-tea experiment, if H_0 was true, the lady did not by sheer chance make correct responses; rather, the experimenter happened by chance to assign cups of tea that matched the responses that she would have made at that time, in any case. It is not particularly useful to have evidence that a person is not making responses at random but it is useful to get evidence that a person can make certain types of discriminations.

The concept of chance performance plays an important part in experiments on extrasensory perception and sometimes leads to experiments which do not test what they are intended to test.

Example 10.4

An experiment is conducted to determine whether an alleged mind reader can read the mind of another person, a "sender." Elaborate precautions are taken to ensure that the two people do not know each other and have no prearranged signals or other ways of faking telepathic transmission. The sender is given five cards and asked to concentrate on one of the cards, whichever one he wants; at the same time, the mind reader, who is isolated from the sender, is asked to specify the card on which the sender is concentrating. One trial each is given for geometrical designs, colors, numbers, and various other stimuli, there being five alternatives for each trial. Suppose that in every one of 10 trials the mind reader selected the card on which the sender was concentrating. Then the probability of this degree of agreement resulting by chance is computed as $(1/5)^{10}$, or about one chance in 10 million. Is that not strong evidence that the mind reader is reading the mind of the sender, since alternative means of transmission have been rendered impossible? No, it is not. It *is* strong evidence that the relationship between the card the mind reader picks and the card the sender observes is not chance. However, maybe all the mind reader is demonstrating is that he tends to think like the sender. After all, there have been numerous demonstrations to show the predictability of responses of people in general when they are asked to name a color, to give a number between 1 and 10, or to perform some similar task. Instead of the answers being uniformly distributed over all possible answers, there tends to be a concentration on one or two popular answers. We might obtain a very small P-value simply because of similarities between the sender and the mind reader. The sender and the mind reader probably are not only members of the same species but members of the same society with enough experiences in common to give them similar preferences. An experiment where the sender, as well as the mind reader, indicates preferences may mislead us into thinking we have evidence of telepathy or mind reading because of the similarity of the preferences of the two people.

To get evidence of mind reading it is necessary to have the chance element in the experiment be introduced by the experimenter. The experimenter randomly selects one of the five cards each time and then has the sender concentrate on it. Under those conditions, the computation of the probability of such a large number of correct responses by the mind reader would test the null hypothesis that the mind reader's response was the same as it would have been if the sender had concentrated on any of the other cards.

10.8   EFFECTS OF TRENDS

Examination of the 12 performance measurements in Example 10.1 shows no tendency for the later measurements to be larger, a tendency, for example, that is often associated with practice effects in the early stages of learning. In experiments where practice effects are likely, it is best to give the subject enough practice trials prior to the test trials for the data to stabilize. However, if there had been a consistent upward or downward trend in the measurements, the test still would have been valid because the random assignment of treatment times to treatments randomizes any beneficial or detrimental effects on performance associated with the treatment time, and the randomization test takes this randomization into account. A consistent upward or downward trend over the duration of an experiment, however, can make a statistical test less powerful (less sensitive) than it would be otherwise, even when it does not affect the validity of the test. Experiments extended over a considerable period of time are especially susceptible to the influence of long-term trends. Let us examine the details of an experiment with a subject where it is necessary to introduce experimental treatments repeatedly over a fairly long period of time.

Example 10.5

A physician wants to find out whether a certain drug will help a patient. From 10 consecutive Mondays, five are randomly drawn and are used as the dates for the administration of the drug. The remaining five Mondays are designated as days on which a placebo is given instead of the drug.

The physician expects the drug to manifest its effect about a day after its administration, and so he takes his physiological measurements of treatment effect on the 10 Tuesdays following the Mondays used in the experiment. If an independent t test was used on the data, with significance determined by data permutation, the procedure would not be basically different from that in Example 10.1, where the experiment was not extended over a very long period of time. The physician, however, does not analyze only those 10 Tuesday measurements. He takes the same kind of measurements on the Sundays preceding the 10 Mondays used in the experiment and derives 10 "change measurements" from the 10 Sunday-to-Tuesday time spans enclosing the Mondays. The measurement on a Sunday is subtracted from the measurement on the following Tuesday to give a change measurement for the enclosed Monday. Then the 10 change measurements are permuted as if they were single measurements, and significance for an independent t test is determined by using Program 4.4.

The reason for using change measurements instead of Tuesday measurements alone is that the physician believed that even without the experimental treatment the subject might consistently improve (or deteriorate) over the 10-week experimental period. The effect of such a consistent trend on the data analysis can be seen by considering what would happen if the measurements for both the placebo and the drug administrations consistently increased over the 10 weeks. Even if the drug tended to produce larger measurements, the upward trend in all measurements could conceal this because the later placebo measurements might be generally higher than the early drug measurements, and such an overlap would render a test based on Tuesday measurements alone relatively insensitive. For a uniform linear trend over the experimental period, the use of Sunday-to-Tuesday change scores controls for the overlap of drug and placebo measurements resulting from the general trend and provides a more sensitive test.

An alternative procedure that can be used to control for general trends over a period of time is to use a time-series test and determine the significance of the test statistic by data permutation. The general objective of time-series analysis when used for testing for treatment effects is the same as that of difference scores: to prevent concealment of treatment effects resulting from a general trend in the data that is not a function of the treatments.

10.9 CARRY-OVER EFFECTS

Parametric significance tables assume not only random sampling from a population of sampling units but also that each sampling unit has a measurement value associated with it that is independent of the constitution of the sample in which it appears. As indicated in Section 10.3, the independence assumption would be unrealistic for random samples of treatment times. The measurement value associated with a randomly selected treatment time ordinarily would be dependent on the sample in which it appeared, because the sample composition determines how many treatment administrations precede any particular treatment time and thus bears on how practiced, fatigued, or bored the subject will be by that time.

The effect of the number of treatment administrations that precede a particular administration can be regarded as a kind of carry-over effect from previous treatments. The validity of determining significance by means of randomization tests for single-subject experiments is unaffected by such a carry-over effect because the effect is constant over all data permutations: for single-subject randomization tests, all data permutations involve the same set of treatment times, and so the number of treatment adminstrations preceding any treatment time is fixed rather than randomly determined.

This carry-over effect is similar to the effect discussed in Example 10.5 and can have a similar depressing effect on the sensitivity of a test unless it is taken into consideration by using change measurements or some other procedure, like time-series analysis, with significance determined by data permutation, to control for the trend induced by the carry-over effect.

The carry-over effect just described should not be considered a treatment effect because its existence does not at all depend on differences between the effects of treatments. The carry-over effect could occur even if the same treatment were given repeatedly. There is another type of carry-over effect, however, which should be considered a treatment effect. That is the *differential* carry-over effect, wherein the subject's response to a treatment at time t_k depends not only on how many treatments preceded that treatment but also on the particular treatments that were given at each of the preceding treatment times. For example, if at time t_k the response to treatment B is influenced by whether the immediately preceding treatment was A or another B treatment, there is a differential treatment effect. Why is this type of effect considered a treatment effect? Because there is a differential effect of the treatments on the measurements. If A and B were identical treatments given different labels, such carry-over effects would not occur, and so differential carry-over effects indicate differential effects of the treatments on the measurements.

What we usually mean by a treatment effect is quite different from a differential carry-over effect. If at treatment time t_k the response to A was different from what it would have been to B, there was a treatment effect, and if the A or B response was the same as it would have been for any possible pattern of preceding treatment assignments, the treatment effect would *not* be a differential carry-over effect. It is important to distinguish this kind of treatment effect from a differential carry-over effect, because both types of effects can occur when a subject takes a series of treatments and can affect the significance of the results. For single-subject experiments, the H_0 of no treatment effect tested by a randomization test is that the association between treatment times and measurements is the same as it would have been for any alternative assignment. If the time-ordered series of measurements associated with the treatment times would have been different for certain possible assignments of treatment times to treatments, H_0 is false; there is a treatment effect. But rejection of H_0 does not imply that there is a treatment effect that is not a differential carry-over effect. The only way of drawing statistical inferences about treatment effects that are not carry-over effects is to use an independent-groups, multiple-subject design, where each subject takes only one treatment. Only in that situation can statistical inferences be drawn about difference in effectiveness of treatments that are not a function of carry-over effects.

The fact that the statistical inference does not necessarily imply a treatment effect for temporally isolated or once-given treatments is not as important as it first appears. Useful generalizations from experiments almost always have a nonstatistical basis because the statistical inferences concern specific subjects at specific times in specific situations, etc., and therefore have little generality. The statistical test for a single-subject experiment therefore can be run to determine whether there is a treatment effect, and then on the basis of theoretical and other nonstatistical considerations the experimenter can decide whether there is sufficient evidence for inferring that there was a treatment effect not attributable to differential carry-over alone. If the experimenter is especially concerned about controlling for differential carry-over effects and cannot do so by using a multiple-subject, independent-groups experiment, he may exercise various experimental controls. If carry-over can result from memory of previous tasks, the experimenter can introduce distracting stimuli between treatment sessions to impair that memory. Or if carry-over could occur through physiological changes of short duration, the experimenter may decide to space the treatment sessions far

apart to cut off this source of differential carry-over. Whatever the situation, it is up to the experimenter to exercise
experimental control if the wants to minimize the influence of
differential carry-over effects.

The possibility that significant results are a function of differential carry-over effects must be considered in multiple-subject experiments, also, when they are of the repeated-measures
kind. Thus, for any experiment where there is more than one
treatment for a subject, an effort should be made to minimize
the likelihood of differential carry-over effects in order to make
the interpretation of significant results less ambiguous.

10.10   REPEATED-MEASURES ANOVA AND THE
CORRELATED t TEST

In multiple-subject experiments, the same subjects are sometimes
given two or more treatments and the data are analyzed by pairing the measurements by subject. The sensitivity of a single-
subject test can be increased by a similar use of pairing. In
single-subject experiments, the variability over the time during
which the experiment is conducted may be considerable in some
cases, and the pairing or clustering of treatment times may be
used to control for the intersessional variability.

Example 10.6

An experimenter has 20 treatment times spread over 10 days.
He performs the following random assignment and pairing. First
he specifies two times for each day, one in the morning and one
in the afternoon. Thus each of the days provides a pair of
treatment times. On each of the 10 days both treatments are
given, the determination of which of the two treatments is to be
given in the morning and which in the afternoon being determined by the toss of a coin. The response measurements are
paired within days, so that there are 10 pairs of measurements
where one member of each pair is a measurement for treatment
A and one member is a measurement for treatment B. Program
5.1, the program for repeated-measures ANOVA and the correlated t test with the significance determined by systematic data
permutation, with ΣT^2 (the sum of the squares of the treatment
totals) as a test statistic, can be used for determining significance for a two-tailed test. If treatment A is predicted to provide the larger measurements, T_A can be used as the one-tailed
test statistic with Program 5.1.

10.11 CORRELATION

A test for correlation with significance determined by a randomization test may be applied to single-subject data whether the
independent and the dependent variables are raw measurements,
ranks, or categorical variables with levels expressed as numbers,
as when 0 and 1 represent two categories of a dichotomy.

Example 10.7

An investigator wants to determine whether a certain drug has
an effect on a patient's blood pressure. He selects six dosage
levels and decides on six times for the drug administration.
Then the six treatment times are assigned randomly to the six
different levels of drug dosage. The time lapse between drug
administration and measurement of blood pressure is set in advance also and is constant over drug dosages. Then the drug
is administered on the six occasions and the six measurements of
blood pressure are obtained. The experimenter predicts that
increases in drug dosage will raise the blood pressure, and so
the one-tailed correlation test statistic ΣXY is computed for all
6! = 720 data permutations, to determine significance.

10.12 FACTORIAL EXPERIMENTS

An example will now be provided to show how a factorial experiment involving only one subject can be devised and the data
analyzed. In this particular example, only two levels of two
factors will be considered, but the procedure followed in the
analysis can be generalized to any number of levels of any number of factors.

Example 10.8

Two factors, illumination and sound, each with two levels, are
investigated to determine their effects on physical strength.
Twenty treatment times are used. Ten treatment times are randomly drawn and assigned to the "light" condition, with the remaining 10 going to the "dark" condition. Then the 10 treatment
times for each of these treatments are randomly divided into five
for a "quiet" condition and five for a "noise" condition. The
sound and illumination conditions are given in four combinations:
quiet and light, quiet and dark, noise and light, and noise and
dark, with five measurements for each treatment combination.

TABLE 1   Factorial Experimental Data

	Quiet		Noise	
Light	11	14	16	22
	12	14	18	24
	15		20	
Dark	15	8	12	3
	8	6	4	12
	14		5	

The measurements in Table 1 show the outcome of a hypothetical experiment, where measures of strength are expressed in arbitrary units. First, consider the test for determining whether the sound variable had any effect on strength within the light condition. This requires a comparison of the two upper cells. The null hypothesis is that the measurement for each of the treatment times in the light condition is independent of the assignment of the treatment time to level of noise. In other words, H_0 is that the presence or absence of noise had no effect on physical strength, for the light condition. Thus, we divide the 10 measurements in the upper row in every possible way into five for the upper left cell and five for the upper right cell, and we compute independent t for each division. Program 4.4 can be used to determine significance by systematic data permutation. For these particular results, a test is unnecessary because it can be seen that the two cells differ maximally, and so the P-value for the obtained results would be 1/252 for a one-tailed independent t test, or 2/252 for a two-tailed t test, since there are $10!/5!5! = 252$ permutations of the measurements to be considered.

Similar comparisons could, of course, be made to determine the effect of sound within the dark condition, the effect of illumination within the quiet condition, or the effect of illumination within the noise condition. For each comparison, the significance could be determined by Program 4.4.

In addition to evaluating the main effects of the illumination and noise factors within particular levels of the other factor, one also can test the main effect of illumination or noise over all levels. Consider how the test would be conducted to test the illumination effect over both quiet and noise conditions. Within each of the two sound conditions, the measurements are divided between the light and dark conditions to test the H_0 of no differential effect of the illumination conditions on strength. The number of permutations within the two levels of sound between the two illumination conditions will be $10!/5!5! \times 10!/5!5! = 63,504$, since each division of the 10 measurements for the quiet condition between the two illumination conditions can be paired with each of the 252 divisions of the 10 measurements for the noise condition. Program 6.1 can be used to test the overall main effect of illumination.

10.13 OPERANT RESEARCH AND TREATMENT BLOCKS

Pavlov repeatedly associated food with the sound of a bell to train a dog to salivate at the sound of the bell. Such conditioning has come to be called *classical conditioning.* Operant conditioning, on the other hand, is conditioning in which reinforcement of a certain type of behavior (e.g., by giving food or shock) leads to an increase (or decrease) in the frequency of that type of response. For example, receiving food for pressing a bar increases the frequency of bar-pressing by a rat, and receiving a shock when stepping onto an electric grid causes a rat to avoid repetition of that behavior.

Behavior modification is applied operant conditioning which is used mainly for getting rid of "problem behavior" in people. The elimination of bed-wetting, thumb-sucking, and stuttering are examples of problems that have been approached through behavior modification.

Operant research frequently involves the intensive study of the individual animal or person. Often the dependent variable is the frequency of response of a certain kind within a period of time during which a certain kind of reinforcement is in effect. (We will call the time period for which the number of responses is recorded a *treatment block.*) For example, a child who is shy is observed over 20 five-minute periods, during each of which the number of times he speaks to another child is recorded. During some of the periods he is given a coin whenever he talks to another child, and in some periods there is no such reinforcement. There would be 20 treatment blocks, and for each block there would be a measurement consisting of the number of times the child conversed with other children during that time interval. The use of treatment blocks, in which a particular treatment is continuously in effect and in which responses of a certain kind are counted to provide a single measurement for that block, does not necessitate special procedures of statistical analysis. The problems of analysis of operant data are not fundamentally different from those of the analysis of single-subject data where there are no treatment blocks, as will be shown in the following discussion of operant research designs. Randomization tests for operant experiments have been discussed elsewhere by Edgington (1984), Kazdin (1976), and Levin et al. (1978).

10.14   ABAB DESIGN

Operant research with two treatment conditions, A and B, is sometimes carried out with a series of treatment blocks with the treatments systematically alternating from block to block. (Treatment A may be the reinforcement condition and B the control or nonreinforcement condition.) This is sometimes called the ABAB design, although it could equally well be called the BABA design. This design controls rather well for effects of practice, fatigue, and other variables that would be expected to systematically increase in magnitude over the treatment blocks, but not for environmental or intraorganismic periodic fluctuations that happen to coincide with alternations in treatments. Careful determination of the length of the treatment block and control over environmental conditions may make it rather plausible that the possibility of cyclical influences with differential effects on the treatment blocks for the two treatments has been minimized. However, the way to render an experimental design invulnerable to objections regarding the preferential postion that one treatment may hold in a series of treatment blocks is to use random assignment.

The restriction that treatments alternate from one treatment block to the next severely limits the possibilities of random assignment. The toss of a coin can determine which treatment is given first, but after that the sequence is completely determined. Consequently, the significance value for any test statistic cannot be less than 0.50 by data permutation, for the ABAB design.

10.15   RANDOM ASSIGNMENT OF TREATMENT
BLOCKS TO TREATMENTS

Like specific treatment times, time intervals (treatment blocks) can be assigned randomly to treatments.

Example 10.9

A child is observed for a period of 1 hour, with 12 five-minute treatment blocks used for recording the frequency of some type of undesirable behavior. Treatment A blocks are intervals of time during which there is no special intervention by the experimenter; they are the control blocks. Treatment B blocks are blocks where the experimenter introduces scolding or some other negative reinforcement expected to reduce the incidence of the undesirable behavior. Consequently a one-tailed test is appropriate. Six of the 12 treatment blocks are selected randomly for treatment A and the remaining six are assigned to treatment B. The number of possible assignments of treatment blocks to treatments is 12!/6!6! = 924. The observed results are as follows:

Treatment block	1	2	3	4	5	6	7	8	9	10	11	12
Measurements	15	12	8	10	7	6	8	5	10	11	7	3
Treatments	A	A	B	A	B	B	A	B	A	A	B	B

The null hypothesis is that for each treatment block the measurement is the same as it would have been for any alternative assignment; i.e., the measurements associated with the treatment blocks are independent of the way in which the treatments are assigned to blocks. The 924 data permutations are the 924 distinctive associations of six A's and six B's with the above sequence of 12 measurements. Treatment A was predicted to provide greater frequencies (larger measurements), and so T_A can be used as the one-tailed test statistic. For the obtained data, $T_A = 15 + 12 + 10 + 8 + 10 + 11 = 66$. The proportion of the 924 data permutations with as large a value of T_A as 66 is the significance value. Determining the significance of T_A or the two-tailed test statistic, $\Sigma(T^2/n)$, can be accomplished by using Program 4.4.

    The test carried out for the 12 treatment blocks is the same as if there were 12 treatment times and a single measurement was made for each time. That the "treatment time" is extended over a 5-minute interval and that the "measurement" for a treatment time is the total number of responses of a certain kind during the 5-minute interval is irrelevant to the determination of significance. The random assignment of treatments to treatment blocks corresponds to the random assignment of subjects to

treatments in a multiple-subject experiment. Consequently the
characteristics of data for which the significance tables for t
provide close approximations to the randomization test significance value for multiple-subject experiments are those for which
the tables provide close approximations for single-subject experiments of this kind. Some of those characteristics are: large
samples, equal sample sizes, relatively continuous data (in con-
trast, for example, to dichotomous data), and absence of outlier
scores.

10.16 COUNTERBALANCED DESIGNS

The completely randomized design we have just considered in
Example 10.9 allows a perfectly valid determination of the significance of the results and would appear to be quite appropriate
for operant research in general. However, if a researcher expects such a strong consistent upward (or downward) trend in
response frequency over time that some direct control over such
a trend is desirable, he may prefer a design like the following
one, in which there is separate random assignment within the
first and second halves of the treatment block sequence.

Example 10.10

Given 12 treatment blocks, of which six are to be assigned to A
and six to B, the assignment can be carried out separately for
the first six and the second six blocks. From the first six
blocks, randomly select three for treatment A and assign the
remaining three to treatment B. Then do the same from the
second six blocks. The number of possible assignments is
$(6!/3!3!) \times (6!/3!3!) = 400$, and each assignment will have half
of the treatment blocks for each treatment in the first half of
the series of blocks and half in the second half of the series.
That is, for each of the 400 possible assignments, three A's and
three B's are in the first half of the series of blocks and three
A's and three B's are in the second half. The data permutation
procedure to determine significance would, of course, be based
on those 400 possible assignments, rather than on the 924
possible assignments in Example 10.9. Program 6.1 can be used
to provide 1-tailed or 2-tailed P-values.

Example 10.10 showed how to perform random assignment
with some degree of control over the effect of a systematic trend
in the data over time. The following example will show how to

exercise even stronger control by restricting the assignments to those that provide a symmetrical series: a series of treatment blocks where the sequential order of treatments from the first to the last is the same as the order from the last to the first.

Example 10.11

Let us again consider 12 treatment blocks, of which six are to be assigned to A and six to B. The treatment blocks are numbered 1 to 12 in temporal order. We now pair the treatment blocks so that each block is paired with a block that is the same distance from the opposite end of the series. The first block is paired with the twelfth, the second with the eleventh, and so on, giving the following six pairs of blocks: 1—12, 2—11, 3—10, 4—9, 5—8, and 6—7. We randomly select three of these pairs for treatment A and the remaining three pairs are assigned to treatment B. The three pairs randomly selected for treatment A are 1—12, 3—10, and 6—7. The remaining three pairs, 2—11, 4—9, and 5—8 are assigned to treatment B. The following data are obtained:

Treatment block	1	2	3	4	5	6	7	8	9	10	11	12
Measurements	5	9	6	8	7	8	5	9	4	7	3	6
Treatments	A	B	A	B	B	A	A	B	B	A	B	A

Neither treatment is given earlier than the other on the average. The two-tailed test statistic can be $\Sigma(T^2/n)$, for which the obtained value is $(37)^2/6 + (40)^2/6 = 494.83$, and the one-tailed test statistic could be T_L, the total of the measurements for the treatment expected to provide the larger measurements. For example, if B was expected to provide the larger measurements, we would compute $T_B = 40$ as the one-tailed test statistic value for the obtained results. There are $6!/3!3! = 20$ possible assignments and 20 associated data permutations. For each of the 20 data permutations, the one- or two-tailed test statistic is computed and the proportion of the data permutations with as large a value of the test statistic as the value for the observed results is the P-value associated with the results.

Since blocks are assigned in pairs to treatments, the six pairs of blocks are the six experimental units. Under H_0, the

sum of measurements for any of the experimental units is independent of the treatment to which the unit is assigned, the sum
being the same for either of the possible treatment assignments.
Thus, for data permutation, we have the following sums associated with the six pairs of treatment blocks:

Pairs	1–12	2–11	3–10	4–9	5–8	6–7
Sums	11	12	13	12	16	13

The 20 possible assignments are the 20 different divisions of
the pairs into three pairs for treatment A and three pairs for
treatment B. Thus the 20 associated permutations of the six
sums shown above can be used to determine significance of T_B
or $\Sigma(T^2/n)$. The use of sums for pairs instead of separate
measurements for each block in performing data permutation permits the use of Program 4.4, the program for the independent
t test with significance based on systematic permutation.

With only 20 data permutations, the smallest possible P-value
is 1/20, or 0.05, whereas the random assignment procedure of
Example 10.9 provides 924 data permutations for the same number of A and B treatment blocks and allows for the possibility
of a P-value as small as 1/924, or about 0.001. If, in fact,
there is no monotonic trend in the data, the use of random
assignment within halves or of random assignment to ensure
symmetry can severely restrict the power of the test. Thus, as
in the case of counterbalancing with multiple-subject designs,
restricting the possible random assignments to provide control
over the serial order of treatments should not be done unless a
strong trend is expected.

10.17 RANDOMIZATION TESTS FOR TREATMENT INTERVENTION

Sometimes in behavior modification studies an experimenter will
select one of the treatment blocks as the time to introduce an
experimental treatment which then remains in effect until the end
of the experiment. Evidence of a treatment effect involves a
comparison of measurements from blocks before *treatment intervention* (introduction of the experimental treatment) with those
blocks following treatment intervention. This comparison is
sometimes made through examining a graph or sometimes it is
based on computation. Occasionally, t or some conventional test

statistic is computed, but frequently there is no attempt to conduct a statistical test of any kind. For the determination of significance by a randomization test procedure the intervention must be random, but random intervention is rare. More commonly, an investigator introduces an experimental treatment when he thinks that the baseline or control observations are stable. But whether the determination of stability is subjective or objective, there is the possibility that use of the baseline stability criterion leads to the introduction of the treatment at a time when the level of the responses would have shown a sharp change even without intervention. To argue on empirical grounds that such a possibility is unlikely to lead to bias in a particular experiment is like arguing in a multiple-subject experiment that a certain systematic assignment of subjects is in fact as good as random assignment; it may persuade a few persons that there is no bias, but it does not provide a sound basis for performing a statistical test. It must be appreciated, therefore, that in the following description of the application of a randomization test procedure, random intervention is included because it is necessary for a statistical test and not because it is in any sense typical. What is to be described is a way to conduct an intervention experiment and analyze the data that will provide a valid determination of the significance of the treatment effect.

Example 10.12

An experimenter decides to test the effectiveness of an experimental treatment (a method of reinforcing behavior) for increasing the frequency of a certain type of desired behavior. He specifies 20 five-minute treatment blocks, for each of which a count of instances of the desired behavior is recorded. When the experimental treatment, which is a method designed to stimulate the desired behavior, is introduced it stays in effect over the rest of the treatment blocks. Thus the intervention of the treatment divides the treatment blocks into two classes: control blocks, which are those prior to treatment intervention, and experimental blocks, which are those after intervention. To prevent the control or experimental samples from being too small, the experimenter randomly introduces the treatment with a sample-size constraint to ensure at least five control and at least five experimental blocks. This sampling constraint is imposed by restricting selection to one of the blocks within the range of blocks 6 to 16. The treatment is introduced at the beginning of the selected block and remains in effect over all subsequent blocks. There are thus 11 possible assignments of the treatment intervention to the blocks. Suppose that the experimenter selected block 8 for the intervention and obtained the following results:

Block	1	2	3	4	5	6	7	8	9	10	11	12	13	14	15	16	17	18	19	20
Data	2	3	4	3	2	3	4	8	9	8	9	10	8	9	9	8	10	9	8	8

The underlined numbers indicate the experimental treatment blocks and the measurements for those blocks. For a one-tailed test where the experimental treatment blocks are expected to provide the larger measurements, $\overline{X}_E - \overline{X}_C$ (the mean of the experimental measurements minus the mean of the control measurements) can be used as a test statistic. The 11 data permutations for which the test statistic is computed are associated with the 11 possible times that could be selected for intervention of the experimental treatment: the beginning of any one of blocks 6 to 16. The test statistic value for the *first* data permutation is the mean of the last 15 measurements minus the mean of the first five, which is $120/15 - 14/5 = 5.20$. For the second data permutation, the test statistic value is $117/14 - 17/6 = 5.52$. The third data permutation is that for the obtained results, which gives an obtained test statistic value of $113/13 - 21/7 = 5.69$. (It may be more convenient to have the first data permutation associated with the earliest potential point of intervention than associated with the obtained results.) Computation of the remaining test statistic values would show that only the obtained data division provides a test statistic value as large as 5.69, and so the probability value for the results is $1/11$, or 0.091.

A two-tailed test statistic for such a design is the absolute difference between means $|\overline{X}_E - \overline{X}_C|$. For either a one- or a two-tailed test, a special computer program is required for treatment intervention designs because of the unusual method of random assignment, which dictates a special method of permuting the data. The data permutation procedure, however, is simple enough to make the development of either a systematic or a random data permutation program fairly easy.

The two test statistics $\overline{X}_E - \overline{X}_C$ and $|\overline{X}_E - \overline{X}_C|$ seem appropriate for detecting treatment effects when there is no general upward or downward trend in measurements over time, that is independent of the treatments. However, $|\overline{X}_E - \overline{X}_C|$ lacks sensitivity in the presence of such trends.

Example 10.13

Let us consider a situation for which the two-tailed test statistic $|\bar{X}_E - \bar{X}_C|$ fails to reflect an apparently substantial treatment effect. Consider the following outcome of a treatment intervention study:

Block	1	2	3	4	5	6	7	8	9	10	11	12	13	14
Data	9	8	7	6	5	4	3	9	8	7	6	5	4	3

Treatment intervention was at the beginning of block 8, and so the measurements for the first seven blocks are control measurements and the underlined measurements, for the last seven blocks, are experimental measurements. The magnitude of the measurements was dropping consistently until treatment intervention, at which time there was a sharp rise although the decline resumed after the rise. The results suggest that fatigue, boredom, or some other measurement depressant caused the measurements to decline in size over time, but that the treatment intervention raised the magnitude of the measurements even though the decline resumed. The two-tailed test statistic $|\bar{X}_E - \bar{X}_C|$ for the obtained results has the smallest possible value, 0, and every other data permutation provides a larger value. Thus, the P-value, instead of reflecting a treatment effect through its smallness, has the largest possible value: 1. If the treatment effect in a treatment intervention experiment is added to a general trend effect, as in this sample, the test statistic $|\bar{X}_E - \bar{X}_C|$ can be quite insensitive to a treatment effect.

Analysis of covariance F, with the block number as a covariate, would seem to be a useful two-tailed test statistic for treatment intervention experiments. (See Winer, 1971, pp. 752−812, for a discussion of analysis of covariance.) By using the block number as a covariate, there is statistical control over temporal trends within treatment conditions. For the data in Example 10.13, where there is a strong tend, the value of F for analysis of covariance is larger for the obtained results than for any of the other data permutations, confirming the general impression given by the data that there was an effect of the experimental treatment. The significance of analysis of covariance F can, of course, be determined by data permutation on the basis of the data permutation procedure described in Example 10.12, with analysis of covariance F computed for each data permutation.

Whether analysis of covariance or some other statistical procedure providing a similar control over trends within treatment
conditions is employed, the significance of the test statistic
should be determined by data permutation because the random
assignment procedure is so different from that assumed by published significance tables as to make the use of such tables
questionable. In particular, the number of possible assignments
is extremely small for the number of experimental units used in
the analysis. For instance, in Example 10.12, where there were
seven control and 13 experimental measurements, there were only
11 assignments that could be made, far short of the 20!/7!13!
possible assignments of seven blocks to the control and 13 to
the experimental treatment associated with random assignment of
treatment blocks to treatments, as described in Example 10.9.
Because of the small number of possible assignments, the treatment intervention experiment is not a very sensitive procedure
and should not be used if any of the more powerful designs
(ones with more possible assignments) that have been discussed
for operant research are applicable.

10.18   RANDOMIZATION TESTS FOR INTERVENTION AND WITHDRAWAL

When there is treatment intervention without withdrawal, there
are only as many data permutations as there are possible intervention blocks, so the randomization test cannot be very sensitive unless there is a large number of blocks. For example,
there must be at least 20 possible blocks for intervention for
the possibility of significance at the 0.05 level. When it is
feasible to have one or two withdrawals of treatment, however,
the sensitivity of a randomization test can be increased substantially.

Example 10.14

An experimenter specifies twenty treatment blocks, for each of
which the total frequency of a certain type of behavior in a
child is to be recorded. Treatment intervention is to be followed by withdrawal, with the following constraints: there will be
at least five control treatment blocks preceding the treatment
intervention, at least five control treatment blocks following
withdrawal of the treatment, and at least five experimental blocks
between the two series of control blocks. The block selected

for intervention restricts the blocks that can be selected for withdrawal. (Intervention or withdrawal will occur at the *beginning* of the selected block.) If the point of intervention was block 6, the point of withdrawal would have to be within blocks 11—16 to satisfy the imposed constraints. But if block 7 was the point of intervention, block 11 could not be a point of withdrawal; only blocks 12—16 would be acceptable as points of withdrawal. In earlier discussions of this example (Edgington, 1975; 1980, pp. 272—274), there was reference to random selection of an intervention block, followed by random selection of a withdrawal block. Ronald Crosier (personal communication, 1982) pointed out that to make all combinations of intervention and withdrawal equally probable, the assignment must involve a *pair* of blocks, the first block being an intervention block and the second a withdrawal block. Thus, the experimenter chooses from the following pairs of blocks, where the first block in a pair is an intervention block and the second a withdrawal block: (6, 11), (6, 12), (6, 13), (6, 14), (6, 15), (6, 16), (7, 12), (7, 13), (7, 14), (7, 15), (7, 16), (8, 13), (8, 14), (8, 15), (8, 16), (9, 14), (9, 15), (9, 16), (10, 15), (10, 16), (11, 16). There are, therefore, 21 equally probable assignments of intervention-withdrawal combinations (pairs), each with a probability of 1/21. (Note that random selection of the intervention block followed by random selection of the withdrawal block would have resulted in *unequal* probabilities for the 21 pairs, ranging from the probability for pair (6, 11), which is 1/6 × 1/6 = 1/36 to the probability for pair (11, 16), which is 1/6 × 1 = 1/6.)

The experimenter randomly selected pair (7, 14), so block 7 was the first block receiving the experimental treatment and block 14 was the first block under the control condition after withdrawal of the experimental treatment. The following results were obtained:

Block	1	2	3	4	5	6	7	8	9	10	11	12	13	14	15	16	17	18	19	20
Data	3	4	4	3	4	3	5	5	6	6	7	5	6	4	3	4	4	3	3	3

The underlined numbers indicate the experimental treatment blocks and the measurements for those blocks. For a one-tailed test where the experimental treatment blocks are expected to provide the larger measurements, the test statistic $\bar{X}_E - \bar{X}_C$ can be used. The obtained value is the mean of the underlined measurements minus the mean of the 13 measurements that precede and follow them, which is $5.71 - 3.46 = 2.25$. The test statistic is computed

for each of the data permutations. The first permutation would
have blocks 6 to 10 for the experimental blocks, the second
would have blocks 6 to 11 for the experimental blocks, and so
on, the last permutation having blocks 11 to 15 for the experimental blocks. There are 21 data permutations for which $\bar{X}_E$ −
$\bar{X}_C$ is computed and, of the 21, only the obtained results provide
a value as large as 2.25. Thus the significance value is 1/21, or
about 0.048. $|\bar{X}_E - \bar{X}_C|$ can be used as a two-tailed test statistic.

Treatment intervention with withdrawal allows for more possible
assignments and, therefore, provides more data permutations than
treatment intervention without withdrawal; but there are still not
enough data permutations to permit a sensitive test. Several
treatment interventions and withdrawals, however, can considerably increase the power of treatment intervention experiments.

Irrespective of the practicality or impracticality of designs
where there is intervention plus withdrawal, they serve a useful
function in illustrating the way randomization tests can be employed
with complex random assignment schemes. The opportunity of
conducting valid statistical tests of treatment effects in experiments employing a sequence of random assignments, where each
assignment may substantially restrict the subsequent random
assignment, should be of interest in a number of multiple-subject
as well as single-subject investigations.

10.19   MULTIPLE SCHEDULE EXPERIMENTS

Hersen and Barlow (1976, p. 225) and Kazdin (1976, 1980) pointed
out limitations of single-subject experimental designs in which
treatments are withdrawn during the experiment. One limitation
is the difficulty in assessing the relative effectiveness of treatments, such as certain drugs, which may have enduring carry-
over effects. Another limitation is the ethical objection to with-
drawing a treatment that may be effective. Sometimes, therefore,
it is desirable to employ a single-subject design in which treatment variables are *not* withdrawn, especially if a statistical test
still can be applied to test for treatment effects.

The treatment intervention design in Section 10.17 permits a
valid determination of treatment effect without the necessity of
treatment withdrawal but that design requires a large number of
treatment blocks to ensure a relatively sensitive randomization
test. Multiple schedule designs (Hersen and Barlow, 1976, pp.
253−258) also involve the application of a treatment without

withdrawal and require only a moderate number of treatment blocks for an adequate test. In a multiple schedule experiment there is more than one possible source of a treatment (e.g., reinforcement), and the source alternates during the course of the experiment, with the treatment being almost continuously provided by some source. Alternating the source of treatment permits the effect of the treatment to be distinguished from the effect of its source. For instance, if two persons alternate in providing reinforcement and nonreinforcement, and the frequency of a certain type of behavior tends to be consistently higher under the reinforcing condition, there is strong evidence of an effect of reinforcement. The following example was presented by Edgington (1982).

Example 10.15

A multiple schedule design was used to assess the effect of social reinforcement on the behavior of a claustrophobic patient, (Agras et al., 1969). The patient was put into a small windowless room and asked to remain in the room until she began to feel anxious, at which time she could open the door and leave. The patient was assessed four times a day to determine the time spent in the room. Two therapists (one at a time) were with the patient each day. The reinforcing therapist praised the subject for staying in the room, whereas the nonreinforcing therapist was pleasant but did not praise the patient. The reinforcing roles of the therapists were switched occasionally. The number of seconds spent in the room for days 2—14 are given in Table 2, which shows results pooled within days to provide a single measurement for the reinforcing sessions and a single measurement for the nonreinforcing sessions for each day. (Day 1 did not involve reinforcement, so it was excluded.)

The data in Table 2 can be used to illustrate the application of a randomization test for a multiple schedule experiment. For this purpose, we will hypothesize the following random assignment. Each day the experimenter tosses a coin to determine which of two times of day (morning or afternoon) will be the reinforcing session and then tosses the coin again to determine which time of day will be associated with Therapist 1. Two functions are served by the coin tosses: they randomly associate treatment times with treatment conditions within each day, and they ensure that systematic differences between reinforcing and nonreinforcing sessions are not a function of a particular therapist being systematically associated with a reinforcing condition.

TABLE 2   Time (Seconds) Spent in Room

							Days						
	2	3	4	5	6	7	8	9	10	11	12	13	14
Reinforcement	250	370	460	520	180	250	290	300	320	240	300	440	500
Nonreinforcement	270	300	420	470	150	200	240	250	280	280	290	380	430

Consider a test of this H_0: for each session (morning or after-noon), the amount of time the lady spends in the room is the same as it would have been under the alternative treatment condition. This implies, for example, that on day 2, if the reinforcing conditions had been switched for the morning and afternoon sessions the woman would have spent 270 seconds in the room under the reinforcing condition and 250 seconds under the nonreinforcing condition. Under the null hypothesis, therefore, the measurements in Table 2 for each day are the same as they would have been under any of the $2^{13} = 8,192$ possible assignments of the reinforcing condition to morning or afternoon sessions for the 13 days; only the designation of a measurement as being associated with reinforcement or nonreinforcement would have been different. Program 5.1, for repeated-measures designs, can be employed, using as a test statistic T_L, the total of the measurements for the treatment predicted to provide the larger measurements, which in this case is the reinforcement condition. The proportion of the 8,192 data permutations derived by switching measurements within days between reinforcement and nonreinforcement groups that provide as large a total for the reinforcement condition as the obtained total is found, by the use of Program 5.1, to be 0.0032, which is thus the one-tailed P-value.

The test of the effect of reinforcement described in the pre-ceding paragraph did not require the random assignment of ther-apists to time of day that was postulated. One therapist could have been associated systematically with morning sessions and the other with afternoon sessions, and a single toss of a coin then would assign a therapist-plus-treatment-time experimental unit to the reinforcement condition. Using two independent random assign-ments, one being the assignment of time of day to reinforcement condition and the other the assignment of therapists to time of day, however, allows us to carry out a second test, to determine whether there was a therapist effect, that is, a differential effect of the two therapists. The data in Table 2 could be permuted for this test, also. Instead of having the data in rows for "rein-forcement" and "nonreinforcement," the top row could be for Therapist 1 and the bottom row for Therapist 2. If, for example, Therapist 1 was the reinforcing therapist for days 2, 3, 5, and 7, the measurements for those days would be in the same rows as in Table 2, but the values 460 and 420 for day 4 would switch rows and the values 150 and 180 for day 6 also would switch rows. The null hypothesis would be that no matter what effect time of day or the reinforcing condition might have, the measurement for each session is the measurement that would have been obtained

under any alternative assignment of therapists. Program 5.1 could
be applied to perform the 8,192 permutations of measurements within days and determine the significance of the one- or two-tailed
test statistics. The P-values obtained would be those associated
with t or |t| for a correlated t test, determined by the randomiza-
tion test procedure.

10.20   REPORTING RESULTS OF STATISTICAL
TESTS

There can be little doubt of the advisability of following the recom-
mendations given in Chapters 4 and 5 in regard to reporting experimental results when significance is determined for a conventional
statistical test by means of a randomization test for *multiple-subject* experiments. By ensuring that the reader recognizes that
the test is familiar and that only the method of determining significance is novel, the experimenter furthers the reader's receptivity
to and understanding of the statistical analysis.

For *single-subject* experiments, on the other hand, calling
attention to the familiarity of the statistical test in a research
report is more likely to mislead the reader and to make the statistical analysis unacceptable than representing the randomization
test as something quite new. Even before the recent articles con-
demning the use of ANOVA for analyzing single-subject data,
experimenters were told that conventional statistical procedures
were invalid for application to single-subject data and they mis-
takenly continue to associate the invalidity with ANOVA and t tests,
not with the significance tables for F and t. To prevent this un-
justified prejudice against the use of conventional tests with single-
subject data from handicapping the communication of research
results, the experimenter should report the results of his randomization tests without relating his statistical test to conventional
tests. Therefore, instead of reporting that an independent t test
was used, with significance determined by means of a randomization test, and reporting the obtained value of t and the degrees
of freedom, the experimenter should refer to the test throughout
the research report as a randomization test. To understand what
the test is sensitive to, the reader should be told the test statistic; but instead of stating that t was computed for each data permutation, it might, for example, be stated that the sum of the A
treatment measurements was computed for every data permutation
as the test statistic.

The type of random assignment employed in conventional independent-groups and repeated-measures designs with multiple-subject experiments should be familiar to readers, but they cannot be expected to know the type of random assignment that is employed with single-subject experiments; so the experimenter sould describe the random assignment of treatment times to treatments and explain that the randomization test is based on permutations of the data that correspond to the possible assignments, when the null hypothesis of no treatment effect is true. It would also be helpful to explain why a single-subject experiment, or a series of single-subject experiments, was conducted instead of a multiple-subject experiment.

REFERENCES

Agras, W. S., Leitenberg, H., Barlow, D. H., and Thomson, L. E. (1969). Instructions and Reinforcement in the Modification of Neurotic Behavior. *Amer. J. Psychiatry 125,* 1435-1439.

Crosier, R. (1982). Personal communication.

Dukes, W. F. (1965). N = 1. *Psych. Bull. 64,* 74—79.

Edgington, E. S. (1967). Statistical inference from N = 1 experiments. *J. Psychol. 65,* 195—199.

Edgington, E. S. (1975). Randomization tests for one-subject operant experiments. *J. Psychol. 90,* 57—68.

Edgington, E. S. (1980). *Randomization Tests.* Dekker, New York.

Edgington, E. S. (1982). Nonparametric tests for single-subject multiple schedule experiments. *Behav. Assess. 4,* 83—91.

Edgington, E. S. (1984). Statistics and single case analysis. In *Progress in Behavior Modification Volume 16* (Miltersen, R. M. Eisler, and P. M. Miller, eds.) Academic Press, New York.

Fisher, R. A. (1951). *The Design of Experiments* (6th ed.). Hafner, London.

Gentile, J. R., Roden, A. H., and Klein, R. D. (1972). An analysis of variance model for the intrasubject replication design. *J. Appl. Behav. Anal. 5,* 193—198.

Hartmann, D. P. (1974). Forcing square pegs into round holes: some comments on "An analysis-of-variance model for the intrasubject replication design." *J. Appl. Behav. Anal.* 7, 635—638.

Hersen, M., and Barlow, D. H. (1976). (Eds.) *Single-case Experimental Designs: Strategies for Studying Behavior Change.* Pergamon, Oxford.

Kazdin, A. E. (1976). Statistical analysis for single-case experimental designs. In *Single-case Experimental Designs: Strategies for Studying Behavior Change* (M. Hersen and D. H. Barlow, eds.). Pergamon, Oxford.

Kazdin, A. E. (1980). Obstacles in using randomization tests in single-case experimentation. *J. Educ. Statist.* 5, 253—260.

Keselman, H. J. (1974). Concerning the statistical procedures enumerated by Gentile et al: Another perspective. *J. Appl. Behav. Anal.* 7, 643—645.

Levin, J. R., Marascuilo, L. A., and Hubert, L. J. (1978). N = nonparametric randomization tests. In *Single-subject Research: Strategies for Evaluating Change* (T. R. Kratochwill, ed.). Academic Press, New York.

McLeod, R. S., Taylor, D. W., Cohen, A., and Cullen, J. B. (1986). Single patient randomised clinical trial: its use in determining optimal treatment for patient with inflammation of a Kock continent ileostomy reservoir. *Lancet,* Vol. I, March 29, 726—728.

Pitman, E. J. G. (1937). Significance tests which may be applied to samples from any populations. *J. R. Statist. Soc. B.,* 4, 119—130.

Weiss, B., Williams, J. H., Margen, S., Abrams, B., Caan, B., Citron, L. J., Cox, C., McKibben, J., Ogar, D., and Schultz, S. (1980). Behaviorial responses to artificial food colors. *Science, 297,* 1487—1489.

Winer, B. J. (1971). *Statistical Principles in Experimental Design* (2nd ed.). McGraw-Hill, New York.

11

Randomization Tests: Additional Examples

The randomization tests discussed in previous chapters have practical value in determining the significance of treatment effects in various kinds of experiments. The tests serve another function as well: they demonstrate aspects of the randomization test procedure that are useful to know when new randomization tests are developed for special purposes. The tests in this chapter also have both practical and heuristic value. These tests, like the earlier ones, provide the experimenter with greater flexibility in design and analysis.

11.1 RANDOMIZATION TESTS FOR MATCHING

When responses are qualitatively different, contingency chi-square tests can be employed to test the null hypothesis of no differential treatment effect. However, a contingency chi-square test is nondirectional, and the results of the test reflect only the consistency, and not the nature, of the association between types of responses and types of treatments. For instance, a contingency chi-square test would not utilize the specific prediction that treatment A would lead to the death of animals and treatment B to survival, but only the prediction that the treatments would have different effects on mortality. In this case, there is a dichotomous dependent variable, so Fisher's exact test can be used as the desired one-tailed test, but there are no standard procedures for accommodating specificity of prediction when there are more than two categories of response.

Let us now consider an example in which we predict a specific kind of response for each subject and determine the significance of the number of correct predictions.

Example 11.1

We conduct an experiment to see whether the songs of baby English sparrows are influenced by the songs of their mothers. The null hypothesis is that a bird's song is not influenced by the mother. We expect the bird's song to be like that of its mother (foster mother, in this example). We obtain six newly-hatched English sparrows to use as subjects. The "treatments" to which the subjects are assigned are six adult female sparrows to serve as foster mothers to the nestlings. The six adult birds represent six different kinds of sparrows, shown in column 1 of Table 1. A recording is made of the song of each of the six adult birds at the beginning of the experiment, for comparison with the songs of the nestlings at a later date. The six nestlings are assigned randomly to the six foster mothers, with one nestling per adult bird. The first two columns of Table 1 show the resulting assignment, the first baby bird being assigned to the grasshopper sparrow, the second to the tree sparrow, and so on. After the young birds have been with their foster mothers for a certain period of time, a recording is made of the song of each baby bird. An ornithologist is given the set of

TABLE 1   Results of Experiment on Acquisition of Bird Songs

Kind of mother sparrow	Assignment of baby birds	Baby bird with most similar song
Cape Sable	3	3
Chipping	5	5
Field	4	6
Grasshopper	1	1
Ipswich	6	4
Tree	2	2

six songs, along with the set of six prerecorded songs of the
adult birds, and is asked to pair the songs of the young birds
with those of the foster mothers to provide the closest correspondence between the paired songs. The ornithologist pairs
the song of bird 1 with the song of the grasshopper sparrow,
that of bird 2 with the song of the tree sparrow, bird 3's song
with the song of the Cape Sable sparrow, and so on, as shown
in columns 1 and 3. To determine the significance of the results
by the randomization test procedure, we use as a test statistic
the number of times the ornithologist pairs a baby bird with the
bird that raised it. The results are shown in Table 1. There
is agreement between the paired birds in columns 2 and 3 for
four subjects: subjects 1, 2, 3, and 5. Thus, the obtained
test statistic value is 4. In order to determine the P-value, we
must determine how many of the data permutations provide as
many as 4 matches, which can be done without a computer program. Data permutations are derived from the data in Table 1
by permuting the sequence of numbers in column 2 to provide
new pairings of columns 2 and 3. (The permutations of numbers
in column 2 represent possible assignments of baby birds to the
birds in column 1, and under H_0, the pairing of columns 1 and
3 is fixed.) There is only one of the 720 pairings that provides
6 matches (perfect agreement). There are no data permutations
with exactly 5 matches because if 5 pairs match, the sixth also
would have to match. There are 6!/4!2! = 15 combinations of
four numbers from the six numbers in column 2 that can match
four numbers in column 3, and for each of those 15, there is
only one way the remaining two numbers in column 2 could dis-
agree with the remaining two numbers in column 3, so there are
15 pairings providing a test statistic value of 4. Thus, there
are 1 + 15 = 16 data permutations with as large a test statistic
value as 4. So the P-value is 16/720, or about 0.022.

In Example 11.1, the determination of significance was not
difficult, but if there had been, say, 20 possible pairings and
8 matches, it would have been considerably more difficult to
determine the P-value. For such cases a formula may be useful.
A discussion by Riordan (1958) of combinatorial aspects of the
arrangement of n rooks on an n × n chess board with no two
rooks in the same column or row has been used by Gillett (1985)
to deal with matching problems. Gillett provided the following
formula to determine the number of pairings of n things for a
one-to-one pairing with n other things:

$$f(j) = \sum_{i=j}^{n} (-1)^{i-j}\binom{i}{j}(r_i)(n - i)! \tag{11.1}$$

where r_i is the coefficient of x_i in the expansion of $(1 + x)^n$.
For the situation described in Example 11.1, the coefficients
are those of $(1 + x)^6$, which are $r_0 = 1$, $r_1 = 6$, $r_2 = 15$, $r_3 = 20$, $r_4 = 15$, $r_5 = 6$, and $r_6 = 1$. There were 4 matches in 6
pairs of numbers, so we would use formula 11.1 by computing
the number of ways of getting 4, 5 or 6 matches as below:

$$f_6 = (-1)^0\binom{6}{6}(1)(6 - 6)! = 1$$

$$f_5 = (-1)^0\binom{5}{5}(6)(6 - 5)! + (-1)^1\binom{6}{5}(1)(6 - 6)! = 0$$

$$f_4 = (-1)^0\binom{4}{4}(15)(6 - 4)! + (-1)^1\binom{5}{4}(6)(6 - 5)!$$

$$+ (-1)^2\binom{6}{4}(1)(6 - 6)! = 15$$

The results of the formula thus coincide with those given earlier,
also yielding a P-value of 16/720, or about 0.022.

Formula 11.1 is inapplicable when the number of things in the
two sets to be matched are not equal or when more than one
member of one set may be matched with a single member of the
other set. In Example 11.1 there were the same number of
baby birds as foster mothers and the ornithologist was instructed
to pair each baby bird's song with one and only one mother
bird's song, so the application of formula 11.1 was appropriate,
but experiments commonly would not meet the requirements associated with the formula. Usually, more than one subject is
assigned to a treatment, and some categories of response occur
more frequently than others, as in the following example.

Example 11.2

Expecting that the color of illumination under which an animal is
raised will be the animal's preference in adulthood, an experimenter randomly assigns 15 young animals to three cages illum-
inated by red, green, and blue lights, with five animals per
color and keeps them there until they have matured. Then they

are removed from their cages and given a choice of cages with the three colors of light. The color of lighting in the cage the animal chooses is recorded, and the following results are obtained:

		Treatments		
		Red	Blue	Green
	Red	3	1	1
Responses	Blue	1	4	2
	Green	1	0	2

It will be observed that the responses tend to be as predicted. Contingency chi-square for the table is 5.60, and with four degrees of freedom, has a P-value of 0.23 on the basis of the distribution underlying chi-square tables. However, as the contingency chi-square test is nondirectional, a test statistic utilizing the specificity of the matching prediction should be more powerful when the prediction is borne out. The test statistic is the number of matches, and the obtained value is 9. A considerable amount of computation would be involved in determining the total number of the $15!/5!5!5! = 756,756$ data permutations with 9 or more matches. The generation of all data permutations by a computer program would be impractical, but Program 11.1, a program based on random permuting of data, can be used. To use the program the user first provides the number of treatments (3) as input data, then the number of subjects for the first, the second and the third treatments (i.e., 5, 5, 5). "Code values" for the treatment levels to serve as input can be any three distinctive numbers, such as 1, 2, and 3. After reading in the codes, the number of data permutations to be generated is entered. The data are then fed in, using the same code numbers for the data, thus substituting 1 for R, 2 for B, and 3 for G. For each data permutation the data codes are paired with the treatment codes as with Program 9.2 for the correlation trend test. Instead of using ΣXY as a test statistic, however, Program 11.1 uses the number of matches among the 15 pairs of code values and "observations" (represented by 1's, 2's and 3's). Program 11.1 applied to the data in this example, using 10,000 data permutations, gives a P-value of 0.034. This is about 1/6th of the P-value of 0.23 given by the (nondirectional) contingency chi-square test.

PROGRAM 11.1   Matching Test: Random Permutation

```
          DIMENSION DATA(100),CODE(10),NCELL(10)
          TOTX=0
          TOTY=0
C . . . READ IN NUMBER OF GROUPS
          READ(5,)NGRPS
C . . . READ IN NUMBER OF OBSERVATIONS PER CELL
          READ(5,)(NCELL(I),I=1,NGRPS)
          NOBS=0
C . . . READ IN CELL CODES
          READ(5,)(CODE(I),I=1,NGRPS)
          DO 1 I=2,NGRPS
   1      NCELL(I)=NCELL(I-1)+NCELL(I)
          N=NCELL(NGRPS)
C . . . READ IN NUMBER OF PERMUTATIONS
          READ(5,)NPERM
C . . . READ IN DATA
          I=0
   2      CONTINUE
          READ(5,)DATA(I+1)
          I=I+1
          IF(I.LT.N)GO TO 2
          PROB=0.
          DO 5 I=1,NPERM
          NN=0
          XM=0.
          DO 4 J=1,NGRPS
          SUM=0
          N=NN+1
          NN=NCELL(J)
          DO 3 K=N,NN
          IF(I.EQ.1)GO TO 3
          CALL RANDOM_$UNIFORM (R)
          ID=K+R*(NCELL(NGRPS)-K+1)
          X=DATA(K)
          DATA(K)=DATA(ID)
          DATA(ID)=X
   3      IF(DATA(K).EQ.CODE(J))XM=XM+1
   4      IF(I.EQ.1)OBT=XM
   5      IF(XM.GE.OBT)PROB=PROB+1.
          PROB=PROB/NPERM
          WRITE(6,500)PROB
 500      FORMAT(/////,30X,"PROBABILITY=          ",F6.4)
```

Program 11.1 is a random permuting program that can be tested on the following data, for which a test based on systematic data permuting should give a P-value of 1/60:

Treatment Codes	Data Codes
1	1, 1, 2
2	2, 2, 1
3	3, 3, 3

There are 7 matches of treatment codes and data codes, and only 28 of all 9!/3!3!3! = 1,680 data permutations provide as many as 7 matches, so the P-value is 28/1,680, or 1/60.

In the two preceding examples, the matching of songs with songs and colors with colors may have given the impression that some observable characteristic of a treatment must be paired with a response in a matching test. In fact, there need not be an observable characteristic of a treatment to match with responses in order for a matching test to be effective. The characteristic of a treatment that is relevant is the *hypothesized response*. An example will help clarify this point. Suppose that on the basis of reactions of humans, it is predicted that Drug 1 given to dogs will produce inactivity, Drug 2 aggressive behavior, and Drug 3 timidity. If the predictions were correct, it would not be because the behavior matched that of the drugs, since drugs do not exhibit behavior; it would be because the behavior matched the behavior predicted to be produced by the drugs. It is useful to regard matching as the correspondence between a *predicted* and an *observed* response. A test of matching is useful whenever we predict a particular kind of response to a particular kind of treatment because there is then a basis for pairing predicted and obtained responses.

Program 11.1 can be used when there are several treatments, not just two or three, and this versatility permits an experimenter to be more flexible in designing matching experiments. Many treatments should be included in an experiment, when responses to them can be predicted, in order to strengthen the test. For example, in a study by Zajonc, Wilson, and Rajecki (1975) in which chicks were dyed either red or green to determine whether they would be attracted to chicks of the same color, the use of more colors might have been effective. Another example of the use of two treatments when more treatments would have increased the power of the study is Burghardt's (1967) experiment in which horsemeat or worms were fed to baby turtles to determine the effect of early feeding on later food preferences.

11.2   RANDOMIZATION TESTS OF PROXIMITY

Matching tests are more useful in some situations than tests of proximity or nearness of an obtained response to a predicted response. Such a situation occurs when a subject responds to identity or nonidentity, rather than similarity. For example, in predicting that children under stress will go to their mothers instead of their nursemaids, fathers, sisters, etc., it might be expected that if the mother is not available any preference for one of the available persons will not be based on similarity to the mother.

In other situations, there may be good reason to measure the quantitative difference between a predicted and an obtained response. To do so requires specification of a quantitative dimension. For instance, if, in the experiment in Example 11.1, we expected a particular dimension of the birds' songs, such as length of the song, variation in loudness, or some other property, to be imitated, reliance on an ornithologist to give a judgment on overall similarity might be inappropriate. Suppose we were interested in the proximity of the sound intensity of the songs of the baby birds to those of their mothers. Instead of having an expert make an overall judgment on similarity of songs, we might measure the sound intensities of the 12 songs and pair the measurements of the baby bird songs with those of the mother bird songs. Then Program 8.1 is applied to the paired measurements to determine the significance of the correlation between the paired sound intensities. The product-moment correlation coefficient is a measure of proximity of paired measurements, relative to their proximity under alternative data permutations, as is evident from the equivalence of r and ΣD^2, the sum of the squared differences between paired measurements, for a randomization test. This equivalence is easy to demonstrate. For paired values of X and Y, $\Sigma D^2 = \Sigma(X_i - Y_i)^2 = \Sigma X^2 - 2\Sigma XY + \Sigma Y^2$. Over all data permutations (pairings of X and Y values) ΣX^2 and ΣY^2 are constant, so $-2\Sigma XY$ is an equivalent test statistic to ΣD^2. As $-2\Sigma XY$ is a test statistic that is equivalent, although inversely related, to ΣXY, which is equivalent to r, ΣD^2 and r are equivalent test statistics.

Program 8.1 cannot be used when there is more than one subject per treatment. When there are several subjects for each treatment, the correlation trend test program, Program 9.1, can be used.

When there are no measurements to pair, making it impossible to use r as a test statistic, it still may be possible to use ΣD^2 as a test statistic for a randomization test of proximity. An example of this kind will be considered in Example 11.5.

11.3   TESTS BASED ON RANDOM SUBSETS OF TREATMENTS

Tests of matching and of proximity both use test statistics expressing a relationship between the treatment conditions and dependent variable measurements. (That also is true of correlation and trend test statistics.) As a consequence of this relationship, it is possible to make effective use of a small number of treatments by randomly selecting them from a population of treatments and then randomly assigning subjects to the randomly selected treatments.

In Example 11.1, we considered the pairing of the songs of six baby birds with the songs of six mother birds to determine the significance of the number of matches. The test was relatively sensitive for that experiment, but if there had been only three nestlings and three mothers, the smallest possible P-value would have been 1/6, the value associated with perfect matching. By increasing the number of possible assignments, however, with only three nestlings raised by three mothers we can have a sensitive randomization test.

Example 11.3

We conduct an experiment to see whether the songs of baby English sparrows are influenced by the mother birds. The experiment is conducted in the manner described in Example 11.1 with this exception: only three baby birds and three mother birds are involved, the three mother birds being randomly selected from the six listed in column 1 of Table 11.1. The random assignment is performed in the following way: one of the six mother birds listed in column 1 of Table 11.1 is randomly selected to be the foster mother of baby bird 1, one of the remaining five is selected for baby bird 2, and one of the remaining four for baby bird 3. Thus, there are 6 × 5 × 4 = 120 possible

assignments. (This is equivalent to selecting three of the six
adult birds randomly and then assigning the three nestlings
randomly to the three selected adults.) As we are concerned
here with only some of the adult birds shown in Table 1, the
assignments and results of this experiment are shown in a different table, Table 2. The first two columns of Table 2
show the assignment: baby bird 1 was assigned to the grasshopper sparrow, bird 2 to the tree sparrow, and bird 3 to the
Cape Sable sparrow. After listening to the prerecorded songs
of all six mother sparrows and to the songs of the three baby
sparrows, the ornithologist pairs the song of bird 1 with that of
the grasshopper sparrow, bird 2's song with the tree sparrow's
song, and bird 3's song with the song of the Cape Sable sparrow, as shown in Table 2. As the ornithologist paired each
young bird with the foster mother that raised it, the number of
matches is 3. To determine the P-value, the number of matches
is computed for all 120 data permutations, and the proportion of
the data permutations with 3 or more matches is the P-value
associated with the results. Under the null hypothesis, the
relationship between columns 1 and 3 of Table 2 is constant
over all possible assignments. The 120 possible assignments
correspond to the ways in which the numbers 1, 2, and 3 can be
arranged over the six rows in column 2. Out of the 120 possible assignments, only the actual assignment provides as many
as three matches; thus, the P-value is 1/120, or about 0.008.

TABLE 2   Assignment of Baby Birds to
Randomly Selected "Mothers"

Kind of mother sparrow	Assignment of baby birds	Baby bird with most similar song
Cape Sable	3	3
Chipping	—	—
Field	—	—
Grasshopper	1	1
Ipswich	—	—
Tree	2	2

Example 11.3 shows how a single "measurement" (pairing of a baby bird and an adult bird on the basis of similarity of songs) from each of three subjects assigned to three treatments can be sufficient for a sensitive randomization test. By selecting three treatments randomly from six possible treatments, rather than nonrandomly specifying or fixing three treatments to which three subjects could be assigned randomly, the size of the smallest possible P-value is reduced from 1/6 to 1/120.

In order to construct a systematic randomization test program for this type of test, one could use the test statistic of Program 11.1 and change the permuting procedure to one that selects combinations of the treatments, then divides the observations among the treatments for each combination of treatments. A random permuting program could randomly select combinations of treatments then randomly divide observations among the treatments, but a combined systematic-random procedure might be more efficient. For example, all combinations of treatments might be systematically generated and for each combination a number of random divisions of data among the treatments could be performed. (See Section 12.16 for further discussion.)

Sampling of treatment levels functions in the same way for measures of proximity. If we had measures of sound intensity of songs of 10 adult birds from which we randomly select three, the product moment r between the paired intensity levels for baby birds and their foster mothers would be the obtained test statistic, and a systematic test would consider all $10 \times 9 \times 8 = 720$ data permutations consisting of pairing the three response intensity levels in every way with any three of the 10 adult bird intensity levels. The proportion of the 720 r's that were as large as the obtained r would be the P-value.

In the case of either matching or proximity test statistics, the sampling procedure can be effective in providing a much smaller randomization test P-value than otherwise would be possible, both for cases where each subject is assigned to a different treatment and where several subjects are assigned to each treatment.

Although product-moment r for a randomization test is equivalent to ΣD^2 for a randomization test with subjects assigned to all treatments and thus is a measure of relative proximity for a set of paired numerical values, it is not an equivalent test statistic to r for a randomization test based on random subsets of treatments. When proximity is predicted, and a set of treatments is randomly sampled, ΣD^2 instead of r, should be computed from the paired measurements. Example 11.4 describes the performance of a randomization test of this kind.

Example 11.4

A social psychologist interested in imitation in children believes that young children will imitate the rate of performance of a new type of activity performed by a person (model) in front of the child. The psychologist specifies 5 rates of beating a drum which, in beats per minute, are: 20, 30, 45, 70, and 100. A random selection of 3 of these rates provides 30, 45, and 100 as the rates to be used in the experiment, and one of these rates is selected randomly for each of 3 subjects. The model performs in front of one child at a time, beating the drum at the rate appropriate for the child. Following exposure to the model's beating of the drum, the child is given an opportunity to beat a drum. The children's rates of drum beating follow:

Model Rate:	20	30	45	70	100
Child Rate:	—	27	55	—	105

There are $5!/2!3! \times 3! = 60$ data permutations, consisting of all pairings of 27, 55, and 105 with three of the five rates of the model. The value of the test statistic ΣD^2 for the obtained results is $(30 - 27)^2 + (45 - 55)^2 + (100 - 105)^2 = 134$. As none of the other data permutations would provide such a small test statistic value as the obtained results, the P-value is 1/60, or about 0.017.

Random sampling of treatments can be used with proximity measures even when there is simply a measure of difference, as in the following example.

Example 11.5

We want to find out whether birds tend to build nests near the trees in which they were raised. An experiment is carried out within a covered enclosure containing six trees, the most widely separated trees being 120 feet apart. The null hypothesis is that the tree in which a bird builds its nest is independent of the tree in which the bird was raised; that is, the tree in which a bird builds its nest is the tree in which it would have built its nest regardless of the tree in which it was raised. The random

TABLE 3   Distances (in Feet) between Trees Where Birds Were
Raised and Trees Where They Built Their Nests

Trees where birds were raised	Trees where nests were built					
	T_1	T_2	T_3	T_4	T_5	T_6
T_1	0	60	110	120	110	40
T_2	60	0 (S_2)	60	87	91	68
T_3	110	60	0	43	62	97
T_4	120	87	43	0 (S_1)	25	93
T_5	110	91	62	25 (S_3)	0	77
T_6	40	68	97	93	77	0

assignment of S's (baby birds) to treatments (trees in which
the birds will be raised) is carried out in the following way:
one of the six trees is randomly selected for S_1, one of the remaining five trees is selected for S_2, and one of the last four
trees is selected for S_3. Thus there are $6 \times 5 \times 4 = 120$ possible assignments. (This is equivalent to randomly selecting
three of the six trees and then randomly assigning one of the
S's to each of the three trees.) The baby bird, along with its
nest and mother, is placed in the assigned tree. Later, when
the baby bird leaves its nest to build a nest of its own, the
tree in which the nest is built is noted. The test statistic is
ΣD^2, where D is the distance between the tree in which a bird
is raised and the tree in which it builds its nest. Table 3
shows the distances of every tree from every other tree and
shows where each of the S's was raised and where each built its
nest. S_1 and S_2 built nests in the trees in which they were
raised, namely, T_4 and T_2, respectively, and so D_1 and D_2 are
both 0. S_3 was raised in T_5 but built its nest in T_4, which was
25 feet away. So $\Sigma D^2 = 0 + 0 + 625 = 625$. Under H_0, S_2 would
have built its nest in T_2 and S_1 and S_3 would have built in T_4,
no matter where they were raised. In terms of Table 3, H_0 is
that the only variation over the 120 potential assignments would

be in the arrangement of S_1, S_2, and S_3 over the six rows within their respective columns. For example, consider the assignment of S_1 to T_1, S_2 to T_2, and S_3 to T_3. The test statistic value for the data permutation for that assignment is $14,400 + 0 + 1,849 = 16,249$ because, under H_0, S_1 would have been 120 feet from where it was going to build its nest, S_2 would have been assigned to the tree in which it was going to build its nest, and S_3 would have been assigned to a tree that was 43 feet from T_4, where it would build its nest. The test statistic ΣD^2 is computed for each of the 120 data permutations. There are only two data permutations that give as *small* a value of ΣD^2 as the obtained value of 625. That is the data permutation for the actual results and the one for the assignment of S_1 to T_5, S_2 to T_2, and S_3 to T_4, which also gives $\Sigma D^2 = 625$. Thus the probability value associated with the results is 2/120, or about 0.017.

11.4   TESTS OF DIFFERENCES IN COMPOSITION

At times an investigator will be interested in changes in composition, rather than changes in amount, of some compound. In studies designed to detect changes in composition resulting from experimental manipulations, a useful index of the composition is the proportion of a compound entity which each component comprises.

A distinction must be made between the differences in composition which contingency chi-square can test and the differences that call for an alternative procedure, such as the one to be described. For experimental applications, contingency chi-square tests differences between treatments with respect to the proportion of experimental units in various categories. The procedure to be described, on the other hand, is concerned with proportional composition within individual experimental units. For instance, if subjects were independently assigned to different treatments, contingency chi-square could be used to test for a difference in the proportion of subjects that died under the different treatments. But if a number of subjects are randomly assigned to treatments expected to affect their blood composition, contingency chi-square could not be used to test the difference between treatments with respect to the proportion of white blood cells that were lymphocytes. The tests to be described would, however, be applicable. To reiterate, the tests of differences in composition that will be described here concern differences in composition of compounds contained within individual experimental units, not differences in proportion of experimental units in various categories.

Components making proportional contributions to a compound of which they are parts can be of various kinds. In investigations of body fluids, like blood or saliva, the proportions may be the relative volume or weight of various constituents. The measurements also could be given as proportions of a length, as, for example, the proportional length of the upper and lower leg. Thus, the entity whose composition is expressed as a pattern of proportions may be a particular substance, a span of time, a volume, a length, or any other entity that can be divided into its constituents.

The simplest test of a difference in composition of experimental units is a test of a difference between the mean proportion of some component under two treatment conditions. For instance, a sample of apples from a tree is randomly divided between two storage conditions to see the effect of the storage conditions on the moistness of the apples after several months. The proportion of water, by weight, in each apple is determined and used as a measurement for the apple. A t test can then be applied to test the difference between the mean proportions for the apples under the two storage conditions. If a randomization test is desired, Program 4.3 or 4.4 could be used. Since the proportion of an apple that is not water is unity minus the proportion that is water, the value of t for any data permutation would be the same whether the proportions used in the test were proportions of water or proportions of dehydrated material.

With only two dependent variables, the pattern of the two proportions is reflected in the proportions for only one of the variables, so a randomization test for such cases could be applied to the proportions for either variable, using programs from Chapter 4 for independent groups designs. But that cannot be done when there are more than two dependent variables, as the distribution of proportions for one dependent variable does not completely determine the distribution for one of the other dependent variables. An adaptation of the multivariate randomization test procedure given in Section 7.6, however, is applicable. The test statistic for the test is SS_B accumulated over all dependent variables, where the dependent variable values are expressed as z-scores. The z-score transformation prevents the arbitrariness of units of measurement for the different dependent variables from influencing the means and standard deviations and, thus, the P-values. The z-score transformation is unnecessary for comparing patterns of proportions, however, because the

TABLE 4   Proportion of Time Spent
With Each Kind of Toy

	Kind of toy		
Treatment	X	Y	Z
	0.59	0.13	0.28
A	0.54	0.18	0.28
	0.68	0.12	0.20
	0.31	0.50	0.19
B	0.35	0.57	0.08
	0.20	0.65	0.15
	0.03	0.19	0.78
C	0.11	0.28	0.61
	0.19	0.38	0.43

measurements on all dependent variables are in the same units,
e.g., inches, millimeters, ounces, or grams. Therefore, the
test statistic proposed for the test for differences between patterns of proportions is SS_B for the proportions accumulated over
all dependent variables. Example 11.6 illustrates the application
of a randomization test using this statistic.

Example 11.6

Nine subjects are randomly assigned to three treatments, with
three subjects per treatment. The treatments are expected to
provide different patterns of proportions across three dependent
variables, so the test statistic chosen is SS_B added over all dependent variables. The three dependent variables are expressed
in the same unit of measurement, time. As in a study by Ewashen
(1980), who used a randomization test similar to the one used in
this example, the subjects are all given the same amount of time
to play with various toys, and the proportion of time spent with
each kind of toy is recorded for each child. The results are
shown in Table 4. Over all three treatments combined, the

mean proportion for X, Y, and Z are the same, but within treatment groups, the proportions tend not to be evenly distributed over the three types of toys. It will be observed that the children under treatment A spent the largest proportion of time with type X toys and the least with type Y toys. The treatment B children, however, spent the greatest proportion of time with type Y toys and least with type Z, and treatment C children spent the greatest amount of time with type Z and the least with type X toys. There are $9!/3!3!3! = 1,680$ data permutations, consisting of all divisions of the nine rows of proportions between the three groups, with three rows per group. The test statistic is $SS_B(X) + SS_B(Y) + SS_B(Z)$, and for each of the data permutations, the test statistic is computed, and the proportion of the 1,680 data permutations with as large a value as the obtained value is the P-value. There is no overlap between treatments of measurements within any of the three columns for the obtained data, so SS_B is maximum for each of the dependent variables, and the obtained data permutation must have the largest value of $SS_B(X) + SS_B(Y) + SS_B(Z)$. As this is an equal-n design, there are $3! = 6$ data permutations with the maximum value of the test statistic. The P-value thus is 6/1,680, or about 0.004.

When proportions have been computed for several dependent variables, they should be recomputed for a particular pair of dependent variables when there is interest in treatment effects on the relationship between those two variables alone. For a test involving only variables X and Y of Example 11.6, for instance, the X and Y proportions as given should not be used in a test but should be changed. The X and Y proportions should be transformed by dividing each proportion by the sum of that proportion and the proportion for the other dependent variable for a subject. For example, the X and Y values for the first subject would be $0.59/(0.59 + 0.13) = 0.819$, and $0.13/(0.59 + 0.13) = 0.181$. There are two advantages in transforming the dependent variable proportions to make them add up to 1 for the dependent variables involved in the test. One advantage is that the transformation makes the test sensitive primarily to the relative distribution of proportions rather than the differences in general levels of proportions between groups. For instance, if every X_A and Y_A was 0.60 and 0.20, respectively, every X_B and Y_B was 0.33 and 0.11, and every X_B and X_C was 0.45 and 0.15, we would obtain the maximum test statistic value, with a P-value of 6/1,680, whereas with each X and Y proportion determined strictly on the basis of those variables and not influenced by the Z variable, (i.e., adjusting the total X + Y proportion to 1) all 9 values would be 0.75 for X and 0.25 for Y, and give a test statistic value of 0, and a P-value of 1. Without adjustment to a total of 1, the test, by being affected by absolute magnitudes of proportions and not just relative magnitudes, may provide significant results when the relationship between X and Y is not affected, and may fail to provide significant results when the relationship between X and Y varies over treatments, the variation being masked by effects of differences in magnitude of X and Y. A second reason for using adjusted proportions is that the test then is equivalent to a test comparing three sets of proportions for X (or for Y). When the X proportion + Y proportion = 1, $SS_B(X) = SS_B(Y)$, so $SS_B(X) + SS_B(Y)$ is simply twice SS_B for either of the variables. Therefore, we can apply Program 4.1 to the distribution of proportions for either X or Y and get the P-value that would be obtained by using $SS_B(X) + SS_B(Y)$.

11.5 RANDOMIZATION TESTS FOR CENSORED SURVIVAL DATA

Although survival analysis can be defined more broadly, typically it concerns the time to failure of physical mechanisms or the time to death of a biological entity (Miller, 1981). A survival time is said to be censored if it cannot be precisely observed but is known to be "at least" or "at most" a certain length of time. In "right censoring," which we will examine here, the observer may know only that the true survival time for an entity was, for example, more than 15 days but not know how much more. A survival time censored at duration time d is designated as d+ to indicate that it is greater than d; thus, if two animals survived 13 and 18 days under a certain treatment and a third animal was known to be alive after 25 days but more specific survival information was not available, the survival times would be listed as 13, 18, and 25+.

The three types of censoring considered in the following discussion are those designated by Miller (1981) as Type 1, Type 2, and random censoring. Type 1 censoring is that in which an investigator fixes a time after which no further failures or deaths can be observed, as the study will be finished at that time. Consider an example in which cancer patients are randomly assigned to two different types of treatment, with the patients entering the study at various times. The duration of the study is 60 months, with the survival time being the number of months a patient survives after entering the study. Those patients entering the study at the beginning and surviving until the end of the study would have censored survival times designated as 60+, a subject who entered the study 15 months after it began and who was still alive at the end of the study would have a censored survival time of 45+, and so on. Type 2 censoring results from the termination of a study as soon as a preset number of deaths of failures occur. For example, in a medical investigation of the relative effect of two adverse environments on mortality, 50 of 100 animals are randomly assigned to environmental condition A and the remaining 50 animals are assigned to condition B, all 100 animals being placed in their experimental environments at the same time. In order to save time, money, and animals, the study ends after ten animals have died, and the remaining 90 animals at that time are returned to their normal environments. Suppose the tenth animal died on day 45. In that case, we would have 10 uncensored survival times, the largest of which was 45, and 90 censored survival times, all of them designated as 45+. The third type of censoring, random censoring, is that in which factors outside the control of the investigator lead to censored observations. For example, subjects may drop out of a study for financial, educational, or other personal reasons. This type of censoring must be independent of the treatment for a valid test; i.e., a censored observation must be censored at the same value under the alternative treatment(s) when the null hypothesis of no differential effect of treatments on mortality is true.

Censoring of data complicates the estimation procedures involved in parametric statistical testing. There is no way to obtain an unbiased estimate of the mean survival time when there are censored observations. Suppose we use Type 1 censoring in tests of components of an electronic device, start the test of all components simultaneously, and terminate the testing at 1,000 hours. We would have the time of failure of all components that failed within 1,000 hours, but of those that had not failed by that time, we would know only that their survival times were 1,000+ hours. In addition to the impossibility of getting an unbiased estimate of the mean survival time, it also is impossible to get an unbiased estimate of the variance of the population of actual survival times. The usual parametric approach, therefore, is to assume a non-normal distribution of survival times, such as a gamma, an exponential, or a Weibull distribution, with a likelihood function which takes into account the existence of censoring.

Rank tests do not require estimation of population parameters, but censoring complicates those tests also. The complication receiving most consideration is that distributions with censored observations may be impossible to rank completely. Consider
the problem of applying the Mann-Whitney U test to censored
survival distributions. The test statistic is the sum of ranks of
one of the groups, so the joint distribution of A and B scores
must be assigned ranks. Consider the following data, based on
random assignment of four subjects to the A treatment and four
to the B treatment, with the prediction that the A treatment
would provide the larger measurements: A: 10, 14, 16+, 16+;
B: 5, 7, 8+, 9. Although measurements 5 and 7 clearly would
be given ranks 8 and 7, respectively, in ranking from high to
low, the precise rank any of the other six measurements would
have if no observations were censored is unknown. To take
care of such ambiguity, Gehan (1965) proposed the use of an
alternative index of the standing of a measurement relative to
other measurements. This index, which we will call a Gehan
score, is assigned to each measurement M and is defined as the
number of measurements known to be smaller than measurement
M minus the number of measurements known to be larger than
M. The Gehan scores associated with the eight measurements
given above are the ones shown in Table 5. The column of
Gehan scores is the second column minus the third column.

After each measurement has been transformed into a Gehan
score, the sum of the Gehan scores for the treatment predicted
to provide the larger survival values is used as a one-tailed
test statistic for the Gehan test, analogous to using the sum of
ranks as a test statistic for the Mann-Whitney test. The absolute
sum of Gehan scores for a treatment is the two-tailed test statistic. (The absolute sum of Gehan scores is the same for the two
treatments.) As sums of Gehan scores for censored distributions
do not have the same null distribution as sums of ranks, the
Mann-Whitney significance tables cannot be used. Furthermore,
as the test statistic distribution is dependent on the number of
censored observations and the value at which they are censored,
significance tables for the sum of Gehan scores are impractical.
Instead, Gehan proposed use of the asymptotic normality of the
null distribution of the sum of Gehan scores to approximate the
P-value.

The data in Table 5 can be used to show how significance
is determined by use of the Gehan test. For a one-tailed test,
the obtained test statistic value is +12, the sum of the Gehan
scores for treatment A, the treatment predicted to provide the

TABLE 5   Gehan Scores for Eight Measurements

Msmt. (M)	No. less than M	No. more than M	Gehan score
5	0	7	− 7
7	1	6	− 5
8+	2	0	+2
9	2	4	− 2
10	3	3	0
14	4	2	+2
16+	5	0	+5
16+	5	0	+5

larger measurements. The sum +12 is divided by the square root of $(G^2)(mn)/(m + n)(m + n - 1)$, where m and n are the sample sizes, and G^2 is the sum of the squares of all eight Gehan scores. This division gives 1.93, which is used as a z-score for a normal distribution to determine the P-value. The proportion of area under the normal curve greater than z = 1.93 is about 0.027, so the P-value given by the Gehan test is 0.027.

Breslow (1970) generalized the Gehan test to more than two treatments, using a test statistic that is the Gehan score counterpart to the Kruskal-Wallis test statistic. Normal curve approximation procedures are used to approximate the P-values for the generalized Gehan test, also.

In order to conduct a randomization test on censored survival data, one could employ other transformations of measurements than Gehan's scoring procedure, but inasmuch as Gehan scoring has widespread acceptance, applying randomization tests to Gehan scores will be a convenient means of demonstrating the use of randomization tests in the analysis of censored data. We will use Gehan's procedure of transforming the data to provide an alternative to ranking, but will not use Gehan's normal curve approximation to the significance value. Instead, we will determine the P-value on the basis of the permutation distribution that Gehan's asymptotic distribution is intended to approximate. After Gehan scores have been assigned to all of the observations, a randomization test can be applied to the Gehan scores as if they constituted the original data. The null hypothesis for a randomization test for censored data is identity of each experimental unit's measurement, censored or uncensored, under any treatment assignment. To reject the randomization test null hypothesis is to conclude that there was a differential treatment effect on at least some of the units. In order to infer that the differential effect is on the actual survival time of some units rather than on the times at which the observations were censored requires that we make the assumption that the times of censoring are independent of treatment assignment. For random censoring, which is outside the control of the experimenter, it therefore is important to ensure that the experimental unit dropped out of the experiment for reasons unrelated to the particular treatment to which it was assigned.

The Gehan test was applied to data in Table 5 to test the difference between treatments A and B. The corresponding randomization test would use the same test statistic, the sum of the Gehan scores for Treatment A. Instead of basing the significance of the obtained sum of 12+ on a normal curve approximation, however, the exact P-value would be determined directly. Program 4.4 applied to the eight Gehan scores would divide the eight scores into four for A and four for B in all 70 possible ways, compute the value of the test statistic for each division, and determine the proportion of the 70 test statistic values that are greater than or equal to +12. Three of the 70 data permutations (including the obtained data permutation) provide a sum of Gehan scores for treatment A that is as large as +12, so the one-tailed P-value is 3/70, or about 0.043. That is the value that was approximated by the value 0.027 based on the normal distribution.

The generalized Gehan test (Breslow, 1970) can be applied when there are more than two groups. It involves computing the Gehan score for each measurement in each treatment group as with the Gehan test and then computing a test statistic related to the Kruskal-Wallis test statistic. As with the two-sample test, the nonparametric tables cannot be used when there are censored data, so a normal curve approximation is employed. Program 4.1, however, can be directly applied to the distribution of Gehan scores to give the P-value being approximated by the normal curve procedure for the Breslow test.

The preceding discussion concerned tests applied to Gehan scores, but randomization tests can readily be applied to other transformations of the data, provided the transformations are made without respect to the way the data are divided among

treatments. Instead of computing Gehan scores, which are based
on whether an observation is known to be greater or less than
measurement M, we could, for example, compute a measure that
incorporated the amount by which each observation was known
to be greater or less than M. This measure for M involves three
steps for its computation: (1) add the amounts by which any of
the measurements is known to be less than M; (2) add the
amounts by which any of the measurements is known to be greater than M; and (3) subtract the second sum from the first sum.
For example, in Table 5, the scores in the Gehan score column
would become -45, -29, $+4$, -14, etc. A randomization test
could be applied to these modified Gehan scores, dividing them
in all 70 ways between treatments A and B to determine the
P-value by the use of Program 4.4. One could, of course, decide to ignore the distinction between censored and uncensored
observations and apply a randomization test to the recorded survival times, where a + sign has not been added to the censored
times. In fact, when rank tests, like the Mann-Whitney U test
or the Kruskal-Wallis test, are used to test the null hypothesis
of no differential effect of treatments rather than identity of
populations, they too can be applied to the ranks of the data by
treating the censored values as uncensored, and P-values from
published significance tables can be used. (One complication at
times could be the number of ties, which would be great when
a large number of subjects are censored at the maximum value.)
Randomization tests for censored survival data from repeated-
measures and randomized block designs are described in
Edgington and Gore (in press).

REFERENCES

Breslow, N. (1970). A generalized Kruskal-Wallis test for comparing K samples subject to unequal patterns of censorship.
Biometrika, 57, 579–594.

Burghardt, G. M. (1967). The primacy effect of the first feeding experience in the snapping turtle. *Psychonomic Science,
7,* 383–384.

Edgington, E. S., and Gore, A. P. (in press). Randomization
tests for censored survival distributions. *Biometrical Journal.*

Ewashen, I. E. (1980). Effects of hospitalization and fantasy
predisposition on children's play with stress-related toys. M.
Sc. Thesis, Department of Psychology, University of Calgary.

Gehan, E. A. (1965). A generalized Wilcoxon test for comparing arbitrarily singly-censored samples. *Biometrika, 52,* 203–223.

Gillett, R. (1985). The matching paradign: An exact test procedure. *Psychol. Bull., 97,* 106–118.

Miller, R. G. (1981). *Survival Analysis.* Wiley, New York.

Riordan, J. (1958). *An Introduction to Combinatorial Analysis.* Wiley, New York.

Zajonc, R. B., Wilson, W. R., and Rajecki, D. W. (1975). Affiliation and social discrimination produced by brief exposure in day-old domestic chicks. *Animal Behavior, 23,* 131–138.

12

Theory

The computer and the work of Dwass (1957) and others on random data permutation have greatly facilitated the application of randomization tests. Major advances in randomization test theory have been less conspicuous. Chung and Fraser (1958) made a theoretical contribution that has profound consequences for the use of randomization tests, but it has received little attention. The basis of the present chapter is an earlier examination of the implications of the Chung-Fraser view of randomization tests based on reference subsets (Edgington, 1983). Reference subsets are subsets of what we will call *primary reference sets*, which are reference sets consisting of data permutations associated with all possible randomizations (random assignments).

Randomization tests usually are described as though they are based on primary reference sets. For such tests, little theory is needed, as the reference set of data permutations "copies" the set of randomizations. The simplicity of these tests has led to the mistaken notion that the logic of randomization tests is trivial. If only primary reference sets could serve as the basis for randomization tests, the utility of randomization tests would be severely constrained. Only tests of the *general null hypothesis* of no effect of any treatment manipulation could be performed; a *restricted null hypothesis*, which refers to some but not all treatment manipulations, would not be testable. A reference set consisting of data permutations for all possible randomizations is inappropriate for evaluating a restricted null hypothesis because the test statistics for certain data permutations may be confounded by effects of treatments to which the restricted null hypothesis does not refer. For instance, a primary reference set is inappropriate for the null hypothesis of no differential effect of treatments A and B for a completely randomized factorial experiment in which subjects could be assigned to treatments A, B, or C. As complex experiments are designed to permit tests of a number of restricted null hypotheses, randomization tests based on primary reference sets are applicable only to simple experimental designs.

12.1 VALID REFERENCE SETS: SYSTEMATIC DATA PERMUTATION

The reference set is the core of a randomization test; it is the "significance table" to which the obtained test statistic is referred to determine the statistical significance of the experimental results. A reference set of data permutations will be valid if it meets this requirement: *The reference set, under the null hypothesis, is the same as it would have been if any of the alternative randomizations associated with the data permutations in the reference set had been the obtained randomization.* We will call a reference set that meets this requirement a *closed reference set.* Assuming that the same test statistic would be computed, regardless of the data permutation obtained in the experiment, the reference set of data permutations uniquely determines the set of test statistic values used to determine significance. In this chapter we will consider only random assignment procedures of the kind where all possible randomizations are equally probable. For such procedures the proportion of the data permutations in the reference set having test statistic values as large as (or, in some cases, as small as) the value for the obtained data permutation is the P-value.

12.2 PERMUTATION GROUPS

Clearly, from what has been said, the validity of a reference set is not a function of its composition. A reference set is valid because of the validity of the procedure that produced it. For a reference set based on systematic data permutation to be valid, we must follow a procedure of generating reference sets that are closed. If the reference set is not closed, the rank of a data permutation in a reference set may vary according to which data permutation is the obtained data permutation, because of shifting

of reference sets, and there is no way to determine the probability under the null hypothesis of getting an obtained data
permutation that has one of the k largest of the n test statistic
values associated with its reference set.

After indicating the impracticality of randomization tests
based on all randomizations for experiments with m subjects for
one treatment and n for another when m and n are large, Chung
and Fraser (1958) suggested that when samples are large we
could consider the data permutations for a *reference subset*:

> There is no essential reason why all the permutations or
> combinations need be used. Suppose we have some rules
> for permuting the sequence of m + n observations. Succes-
> sive application of these permutations may produce new per-
> mutations but eventually there will be no new permutations
> produced. The resulting collection of different permutations
> of the sequence of m + n observations is then a group.
> (p. 733.)

Note that the group property refers to the procedure for permuting the data, not to the set of data divisions resulting from
application of the procedure. The essence of a *permutation
group* is contained in the following definition: *A permutation
group is a collection of permuting (rearranging) operations that
produce the same set when applied to elements in the set, regardless of the initial element to which the operations are applied.*
A permutation group may be applied either to a data permutation
to generate a set of data permutations (a reference set) or to a
particular randomization to generate a set of randomizations.
For a reference set to be closed requires that it be produced by
a permutation group, but a set produced by a permutation group
is not necessarily closed. It is closed only if the data permutations in the reference set represent outcomes for alternative
assignments (randomizations), under the null hypothesis. A
permutation subgroup is a permutation group but one which
generates only a *subset* of the elements that would be generated
by the more inclusive permutation group.

12.3 DATA-PERMUTING AND RANDOMIZATION-REFERRAL PROCEDURES

The standard way to construct a reference set for a randomization test, which will be called the *data-permuting* procedure, is

the procedure discussed in this book, which involves permuting
the obtained data permutation to produce a reference set without consideration of which assignment (randomization) is associated with the obtained data permutation. It is the usual way of
performing a randomization test, and the validity of the procedure depends on employing a permutation group.

An alternative way to construct a reference set is the *randomization-referral* approach. This approach is of little practical
value but it is useful in revealing the rationale underlying the
data-permuting procedure. To construct a reference set for a
randomization test based on the set of all randomizations, the
entire randomization set, which, of course, contains the obtained
randomization, is transformed into data permutations comprising
the reference set by noting the subject (or other experimental
unit) that provided each measurement in the obtained data permutation. For a randomization-referral test based on a subset
of all randomizations, the set of all randomizations is divided
into subsets in advance and the randomization subset containing
the obtained randomization is transformed into a reference subset.

Kempthorne, in linking the randomization test procedure to
random assignment, has constently described the construction of
a reference set in terms of the randomization-referral approach.
For example, Kempthorne and Doerfler (1969) indicated that if
an experimenter wants to perform a randomization test

> he will superimpose on the data each of the possible plans
> and will obtain C_1, C_2, . . . , C_M, one of which will be
> C_{obs}. He will then count the proportion of M values, C_1,
> C_2, . . . , C_M, which equal or exceed C_{obs}.   (p. 237).

This transformation of "plans" (randomizations) into data permutations is what was designated in the preceding paragraph as
the randomization-referral approach to construction of a reference
set.

Because the data-permuting approach involves permuting the
data without paying attention to the randomization that is associated with each data permutation in the experiment, it is desirable to ensure that no matter which randomization in the reference set was the one obtained, the reference set would be valid.
This can be done by considering how a reference set for a data-
permuting procedure could be generated by a randomization-
referral counterpart that explicitly took into consideration the

identity of the obtained randomization, i.e., the assignment that
was carried out. The validity of a reference set generated by
a randomization-referral procedure is easy to assess; thus, a
tactic that will be employed in subsequent sections of this chapter
to demonstrate the validity of a reference set for a data-permuting test will be to show that the same reference set can be
generated by a valid randomization-referral counterpart to the
data-permuting test.

12.4 REFERENCE SETS FOR GENERAL NULL HYPOTHESES

Consider an experiment to determine the effect of drug dosage
on reaction time. It is predicted that larger dosages will lead
to longer reaction times. Four subjects are randomly assigned
to 15, 20, 25, and 30 mg of a certain drug, one subject per
dosage level. The reaction time measurements associated with
the dosages are 20, 37, 39, and 41, respectively. A randomization test using product-moment r as a test statistic could be conducted by the use of Program 8.1. There would be 24 data permutations, and the highest possible correlation is associated with
the obtained results, so the one-tailed P-value is 1/24, based on
a permutation group producing the primary reference set.

We will now consider how the test would be carried out by
using a reference set derived by a randomization-referral procedure. The entire randomization set of 24 randomizations is
constructed before the experiment, using the letters a, b, c and
d to designate particular subjects. Every possible randomization
can be represented by a particular sequence of the four letters.
The obtained randomization is acbd, which means that subject a
was assigned to 15 mg, c to 20 mg, b to 25 mg, and d to 30 mg.
The obtained randomization, in conjunction with the obtained
data permutation, provides the basis for transforming the randomization set into a set of data permutations comprising the reference set. Under the null hypothesis, the measurement associated
with each subject in the experiment is the measurement that subject would have provided under any of the alternative randomizations. As the measurements for subjects a, c, b and d were
20, 37, 39, and 41, respectively, those numerical values can be
substituted for the respective letters in each of the randomizations to transform a randomization into a data permutation. The
proportion of data permutations in the resulting reference set
that have correlation coefficients as large as 0.87, the value for

the obtained data permutation, is the P-value. The obtained data permutation is the only one of the 24 with such a large correlation; thus, the P-value is 1/24, or about 0.04, the same as for the data-permuting procedure.

Obviously, the reference set for the randomization-referral procedure is valid, and the validity of the data-permuting test can be shown by demonstrating that it has the same reference set and thus necessarily gives the same P-value. The randomizations that could result from the random assignment procedure are the 24 randomizations represented by permutations of the letters a, b, c, and d, and the randomization-referral procedure associates 24 permutations of the measurements 20, 37, 39, and 41 with those randomizations. The data-permuting procedure produces the same reference set by directly, rather than indirectly, generating the data permutations through applying the same permutation group of operations to the obtained data permutation. This permutation group, when applied to the obtained randomization would generate the randomization set that is then transformed into a reference set of data permuations.

For a test of any general null hypothesis on the basis of a primary reference set, a data-permuting procedure will provide a valid reference set when the permutation group applied to the obtained data permutation in generating the reference set is also a group that, if applied to the obtained randomization, would generate all possible randomizations.

12.5 REFERENCE SUBSETS FOR GENERAL NULL HYPOTHESES

Next, consider how to generate a reference subset to test the hypothesis of no effect of drug dosage on reaction time. First we will consider the conduct of a data-permuting test and then justify it by comparing it to the corresponding randomization-referral test.

The rule for permuting the data to get a reference subset is very similar to the one used in an example by Chung and Fraser (1958): Move the first measurement of the obtained data permutation (i.e., 20, 37, 39, 41) to the last position to generate a second data permutation, transform that data permutation in the same way to generate a third data permutation, and so on, until no new data permutations are produced. Application of these operations results in the data permutations shown in Table 1. There are only four data permutations in the reference subset

TABLE 1 Data Permutations and
Associated Correlation Coefficients

Drug dosage (mg)				
15	20	25	30	r
20	37	39	41	0.87
37	39	41	20	−0.66
39	41	20	37	−0.36
41	20	37	39	0.15

shown in Table 1 because transforming the fourth data permutation by moving "41" to fourth place in the sequence gives
the initial data permutation. The collection of operations providing this reference subset thus constitutes a permutation
group. This permutation group produces only a subset of the
data permutations associated with all randomizations and, consequently, is a subgroup of the permutation group that generated all 24 data permutations.

To assess the validity of the reference subset of four data
permutations generated by the data-permuting procedure, generation of the same reference subset by a randomization-referral
procedure will be examined. Before the experiment we consider
the set of all randomizations to determine whether it can be partitioned in such a way that the members of each subset could be
generated by applying to any member the same permutation subgroup that was applied to the obtained data permutation by the
data-permuting procedure. The partition of the randomization
set shown in Table 2 is such a partition. Within each randomization subset all four randomizations can be derived from any
randomization in the subset by moving the first letter of a randomization to fourth position in the sequence to generate a new
randomization, transforming that randomization in the same way,
and so on until no new randomizations are produced. In other
words, we apply the same permutation subgroup to the randomizations within subsets as was applied to the data permutations in
the data-permuting test to generate the reference subset. All
24 randomizations are represented in the partitioned set, and
each is contained within a subset; therefore, the randomization
that will be associated with an obtained data permutation must be

TABLE 2 Partition of a Randomization Set

	Drug dosage (mg)																							
	15	20	25	30	15	20	25	30	15	20	25	30	15	20	25	30	15	20	25	30	15	20	25	30
a	a	b	c	d	a	b	d	c	a	c	b	d	a	c	d	b	a	d	b	c	a	d	c	b
b	b	a	d	c	b	a	c	d	b	d	a	c	b	d	c	a	b	c	a	d	b	c	d	a
c	c	d	a	b	c	d	b	a	c	a	d	b	c	a	b	d	c	b	d	a	c	b	a	d
d	d	c	b	a	d	c	a	b	d	b	c	a	d	b	a	c	d	a	c	b	d	a	b	c

a member of one of the six subsets. Thus, no matter which randomization is associated with an obtained data permutation, the data permutations in the reference set generated by the data-permuting procedure are associated with the randomizations in the subset containing the obtained randomization, and so the same reference set would be used by either the data-permuting or the randomization-referral test. The reference set for the randomization-referral procedure is valid, and since the data-permuting procedure provides the same reference set, that set also is valid.

For a test of any general null hypothesis by use of a reference set or subset, a data-permuting procedure is not valid unless the permutation group applied to the obtained data permutation is a group that would be valid to apply to every randomization with a randomization-referral procedure. If the permutation group applied to the obtained data permutation could provide data permutations associated with impossible randomizations, this condition would not be met. Such would be the case if we tossed a coin to assign subjects a and b to 15 to 20 mg and tossed it again to assign subjects c and d to 25 or 30 mg and the data were permuted by the procedure described above, whereby the first measurement is moved to the end repeatedly to generate four data permutations. If the obtained data permutation had been associated with randomization abdc, the data permutations generated would be those associated with abdc, bdca, dcab, and cabd, but the last three randomizations would be impossible under the assignment procedure.

12.6 REFERENCE SUBSETS FOR RESTRICTED NULL HYPOTHESES

Reference subsets must be used to test restricted null hypotheses because data permutations for certain randomizations are indeterminate. Those data permutations are indeterminate that are associated with randomizations differing from the obtained randomization in the assignment of subjects with respect to a treatment manipulation not specified in the null hypothesis. Two kinds of restricted null hypotheses are: (1) hypotheses restricted to certain treatment conditions, and (2) hypotheses restricted to certain factors. Tests of both types of restricted null hypotheses have been described in earlier chapters, and the following discussion will provide a rationale for those tests.

12.7   REFERENCE SUBSETS FOR PLANNED
AND MULTIPLE COMPARISONS

Suppose six subjects are assigned randomly to three treatments,
with two subjects per treatment, and the following results are
obtained: A: 2, 6; B: 3, 8; C: 10, 11. If the null hypothesis is restricted to the difference between the effects of B
and C, there is no basis for switching measurements for A with
those for B or C in the generation of data permutations since
treatment A may have a different effect. If in fact we permuted
data over all three treatments we would be employing a permutation group, but the primary reference set generated would not
be a closed reference set because there would be no justification
for permuting the A measurements. The reference set would
contain data permutations which, under the restricted null hypothesis, might not be the results that would have been obtained
if the randomizations associated with those data permutations had
been the actual assignment.

The test of the restricted hypothesis can be carried out by
a data-permuting procedure in the following manner. Permute
the four measurements for B and C between B and C with two
measurements per treatment in all $4!/2!2! = 6$ ways, keeping the
assignment of measurements 2 and 6 to A fixed. The resulting
permutation subgroup provides the six data permutations shown
in Table 3, which consitute the reference subset. For each
of the data permutations, the relevant test statistic for reflecting
the expected difference between effects of treatments B and C
is computed and the proportion of the test statistic values that
are as large as the value for the first (obtained) data permutation is the P-value.

The reference subset provided by this data-permuting procedure is the reference subset that would be produced by a particular randomization-referral approach. The latter would use
randomization subsets generated by applying to a randomization
subset member the permutation subgroup that was applied to the
obtained data permutation to generate the reference subset for
the data-permuting procedure. If we knew the obtained randomization, i.e., if we knew the treatment to which each subject was
assigned, and we applied the data-permuting permutation subgroup to it, a subset of six randomizations would be generated.
We could determine before the experiment all possible randomization subsets that could be generated by the permutation subgroup. Those randomization subsets are the ones resulting from
a partitioning of the $6!/2!2!2! = 90$ possible randomizations into

TABLE 3   Reference Set
of Six Data Permutations

Treatments		
A	B	C
2, 6	3, 8	10, 11
2, 6	3, 10	8, 11
2, 6	2, 11	8, 10
2, 6	8, 10	3, 11
2, 6	8, 11	3, 10
2, 6	10, 11	3, 8

15 subsets. Each subset consists of six randomizations where the two subjects for A are the same and the four subjects for B and C are divided in all six ways between B and C. Whatever the obtained randomization may be, it will belong to one of the 15 subsets, and transforming the randomizations in that subset into data permutations will provide the same reference subset as that given by the data-permuting procedure. To ensure the validity of a reference subset for a restricted null hypothesis, we must use only permutation subgroups that generate data permutations that, under the null hypothesis, are derivable from the associated randomizations. Holding the measurements for A fixed and permuting the other four measurements over B and C does provide only data permutations that can be determined from their associated randomizations. The permutation subgroup employed in the data-permuting test provides only data permutations which are associated with possible randomizations and are determinate under the restricted null hypothesis. Thus, the reference subset is valid, and the data-permuting procedure is valid.

More than one data-permuting procedure can be valid for testing a certain restricted null hypothesis. There may be a number of permutation subgroups that, in a given situation, could provide a valid reference subset. For example, for a repeated-measures experiment in which each of n subjects takes four treatments, A, B, C, and D in a random order, a test of no differential effect of treatments A, B, and C could be based on a reference subset consisting of all data permutations under the restricted null hypothesis, of which there are $(3!)^n$, or on a reference subset of 3^n data permutations based on the permutation subgroup that produces only the ABC, BCA, and CAB randomizations of the orders in which any subject would take the treatments.

12.8   REFERENCE SUBSETS FOR FACTORIAL DESIGNS

Another type of restricted hypothesis is one restricted to certain factors in a factorial experimental design. Suppose eight subjects are assigned randomly to any of four cells in a 2 × 2 design, with two subjects per cell, and that the results shown in Table 4 are obtained. A test dependent on all $8!/2!2!2!2!$ divisions of the data among the four cells would provide a primary reference set based on a permutation group, but the set would not be valid because it would include data permutations for randomizations for which, under the restricted null hypothesis, the data conceivably might be different. A valid test of the effect of only some of the factors in a completely randomized experiment can, however, be based on a reference subset. Consider a data-permuting procedure to test the effect of factor A. Without assumptions about the effect of factor B, 36 data permutations can be generated on the basis of the obtained data permutation and the null hypothesis of no effect of factor A. The 36 data permutations are those generated by exchanging the B_1 measurements between A_1 and A_2 in the six possible ways, while also exchanging the B_2 measurements between A_1 and A_2 in the six possible ways. These 36 data permutations constitute the reference subset used for determining the significance of factor A.

TABLE 4   Data from a Factorial Experiment

		Levels of factor A	
		A_1	A_2
Levels of	B_1	3, 5	8, 7
Factor B	B_2	1, 4	10, 12

Suppose we wish to conduct a randomization-referral test that necessarily will give the same reference subset as that provided by the data-permuting procedure. Prior to the experiment, we partition the 2,520 possible randomizations in such a way that one of the randomization subsets will consist of randomizations associated with the data permutations in the reference subset for the data-permuting procedure. The partition can be constructed by applying to each of the 2,520 randomizations the permutation subgroup used in the data-permuting test, resulting in 70 disjoint subsets of 36 randomizations each. Within each of the 70 randomization subsets, the four subjects for level B_1 and the four subjects for level B_2 will be assigned to the same levels of B over all 36 randomizations in the subset. When the members of the subset within which the obtained randomization falls are transformed into data permutations, the resulting reference set is the same as the reference set for the data-permuting test. As the reference set for the randomization-referral test obviously meets the validity requirement, the data-permuting test also must be valid.

12.9 PERMUTING DATA: MULTISTAGE MODEL

A convenient device for remembering how to permute data in testing a restricted null hypothesis is to conceive of the overall random assignment procedure as being made up of consecutive stages (Edgington, 1969, pp. 192—194). It is a way of conceiving of a group of permuting operations being subdivided into subgroups. In a completely randomized two-factor experiment, for example, the overall random assignment procedure can be thought of as being composed of two stages for the purpose of testing the effect of factor B: the first stage is the assignment of subjects to levels of factor A, and the second stage is the assignment of subjects within levels of A to levels of B. The reference subset for testing the effect of factor B is produced by permuting the data in the same manner as they would have been permuted to test the hypothesis if the first stage of assignment had been nonrandom, as in blocking the subjects in a randomized block design. To test the effect of factor A, we can regard the assignment of subjects to levels of A as the second stage. The type of restricted null hypothesis which refers to only some of several treatments in a single-factor design would be treated in an analogous manner. With random assignment of subjects to treatments A, B, and C, to test the hypothesis of identity of B and C effects, we may regard the first stage of

assignment as the assignment of subjects to treatment A or to
a treatment that is not A and the second stage as the assignment to B or C of the subjects not assigned to A. To produce
the reference subset, we divide the data between B and C in
the way the measurements would have been divided to test the
hypothesis if the second stage had been the only random assignment performed in the experiment. To test any restricted null
hypothesis, all stages of assignment of subjects with respect to
treatment variation not specified in the null hypothesis can be
regarded as having been completed in a nonrandom manner before the final stage of assignment. The latter is the part of
the assignment to treatments that is relevant to the null hypothesis and is regarded as the only assignment that is random.

12.10   PERMUTING DATA: DATA-EXCHANGING MODEL

An alternative way of conceiving of the production of a reference
set by the data-permuting procedure is in terms of exchanging
data. In the following discussion of the data-exchanging model,
there will be consideration of the cells among which exchanges
of data are permitted by the null hypothesis, a cell being defined
as the intersection of a level from each of the factors built into
a design. In a completely randomized design, all factors are
manipulated, so each cell represents a distinct treatment condition. With randomized block designs, factors used for class-
ifying experimental units are not experimentally manipulated, so
two or more cells may represent the same treatment condition for
different kinds of subjects. The null hypothesis determines the
cells to which exchanges of data are restricted.

Consider the cells to which subjects may be assigned in an
experiment with four levels of factor A and three levels of factor
B, as shown in Table 5, where the cells are numbered for iden-
tification. Table 5 will be used to evaluate the permuting of
data for single-factor and two-factor experiments and will provide a basis for evaluating more complex designs.

For a single-factor experiment with four treatments, the four
treatments to which subjects can be assigned can be represented
by the four cells for level B_1. The null hypothesis of no effect
of factor A is tested by dividing the data in all possible ways
among those four cells. To test the null hypothesis of no differential effect of the first two levels of A, on the other hand, the
data are divided between only the first two cells.

TABLE 5   Data-exchanging Table

		Levels of Factor A			
		A_1	A_2	A_3	A_4
Levels of	B_1	1	2	3	4
Factor B	B_2	5	6	7	8
	B_3	9	10	11	12

For a two-factor, completely randomized experiment, the hypothesis of no effect of either factor is based on a primary reference set resulting from exchanging data in all possible ways among all 12 cells. The hypothesis of no effect of factor A is tested by permuting data within three sets of cells: {1, 2, 3, and 4}, {5, 6, 7, and 8}, and {9, 10, 11, and 12}. This leads to all possible patterns of data over the 12 cells that can be generated by exchanging data within those three sets of cells. The null hypothesis of no effect of either factor at the lowest level of the other factor is the hypothesis of no differential effect of cells 1, 2, 3, 4, 5, and 9, so the permutation group exchanges data among those cells in all possible ways.

Hypotheses for randomized blocks can be expressed in terms of exchanges of data between cells in a similar fashion. For example, in a randomized block design, the three levels of B could represent three types of subjects or, for a repeated-measures experiment, the three levels of B could refer to three subjects, each of which was subjected to all four levels of treatment A. The null hypothesis of no effect of factor B is not testable by a randomization test when subjects are not assigned randomly to levels of factor B. The main effect of factor A or the effect of A within certain levels of B, however, is testable, and the test can be carried out as for a completely randomized design.

12.11   OPEN REFERENCE SETS: TREATMENT INTERVENTION AND WITHDRAWAL

In single-subject experiments, one might be interested in testing the effect of either intervention or withdrawal, without assumptions

about the effect of the other. When we consider a procedure
which randomly selects a pair of intervention and withdrawal
times, as in Section 10.18, we see some complications from the
standpoint of permutation group theory.

Consider an experiment where the intervention and withdrawal times are selected from only six pairs of treatment blocks:
8−12, 8−13, 8−14, 9−13, 9−14, and 10−14, where the first number indicates the block of time when the treatment was introduced
and the second number the block at the end of which it is withdrawn. The sixth pair of treatment blocks (i.e., 10−14) is
selected, and after treatment intervention and withdrawal, the
data for all treatment blocks in the study are recorded. To test
the effect of treatment intervention without any assumptions
about the effect of withdrawal, we decide to compute our test
statistic, which is the mean of the treatment blocks minus the
mean of the control blocks (the blocks preceding treatment intervention), for each of the six possible randomizations to get our
reference set. Under the null hypothesis of no effect of intervention, the test statistic values for the intervals 8−12, 8−13,
etc. are just what they would have been if any of those pairs
of treatment blocks had been the obtained randomization. Unfortunately, the reference set would not be closed. Only the
sixth randomization allows us to infer, under the null hypothesis,
the test statistic values for all six randomizations. If the first
randomization (i.e., 8−12) had been the obtained randomization,
there would have been only the data permutation for that randomization in the reference set because the other five pairs include
blocks that would be withdrawal blocks if the first randomization (8−12) was the one performed, and there is no basis for
assuming the withdrawal blocks to have no effect when our null
hypothesis is restricted to the difference between intervention
and control conditions. Depending on the pair of treatment
blocks selected for intervention and withdrawal, we would have
a reference set of one, two, three, four, five, or six data permutations. The adverse effect that lack of closure of the reference set could have can be appreciated by considering what
would happen if there was no effect of either intervention or
withdrawal but there was a consistent upward trend in the treatment block measurements as a function of extraneous factors
(practice, motivation, etc.). In that case, each of the six reference sets, which involve comparisons of the obtained test statistic value with only test statistic values for earlier treatment
blocks, would be the largest one in the set. A similar lack of

closure of the reference set would be encountered if we were
to try to test the effect of withdrawal alone. The lack of closure
of the reference sets for the test of intervention or withdrawal
alone means that there is no basis for a valid randomization
test of those specific effects with such an experimental design.

12.12   CLOSED REFERENCE SETS: DROPOUTS

In Section 4.12 it was indicated that when subjects drop out of
experiments without providing measurements, for reasons that
are unrelated to the treatments, one can use either the primary
reference set or a subset of it. Although the discussion in that
section was in terms of the independent t test, the choice of two
alternative reference sets in the case of dropouts is available for
any experimental design. The primary permutation group permutes the data and markers indicating dropouts, so that for some
data permutations the dropout(s) will be associated with certain
treatments and for other data permutations with other treatments.
As the dropouts to which we are referring provide no measurements, the test statistic should take this fact into consideration
in computing means; otherwise, if the sum of the measurements
for a treatment is divided by the total number of subjects taking
the treatment, including the dropout, the dropout is, in effect,
given a score of 0. Nevertheless, if useful test statistics can
be derived for analysis based on a primary reference set, a randomization test can be validly conducted. A reference subset is
an alternative that can be more readily adapted to the use of
conventional randomization test computer programs. The permutation subgroup for generating the reference subset holds the
assignment of dropouts to treatments fixed and permutes the remaining data. The two systematic permuting procedures will not
necessarily give the same P-value, but both are valid procedures.

In Section 11.5, we considered censored survival distributions, in which the "dropouts" provided incomplete information;
we knew that a dropout survived a certain length of time but
not how much longer. Instead of the procedure described in
Section 11.5, it would have been valid to have used a primary
reference set in which the measurements for dropouts were discarded and markers were assigned, as in Section 4.12. Useful
data, however, which by this procedure would be lost, is saved
when a primary reference set based on retaining the measurements
for dropouts is employed.

12.13   OPEN REFERENCE SETS: TRANSFORMATIONS

In the discussions in this book of the invalidity of permuting
residuals for interaction (Section 6.7), and the necessity of computing F for analysis of covariance separately for each data permutation rather than permuting residuals based on the obtained
data permutation, the criticism was not that there was an absence
of a permutation group. Permutation groups were employed but
the reference sets they produced were not those which would
have been produced if some of the randomizations associated with
alternative data permutations had provided the obtained data
permutation; that is, the reference sets were not closed.

12.14   STOCHASTICALLY CLOSED REFERENCE SETS

A reference set for systematic data permutation was described
in Section 12.1 as valid if, under the null hypothesis, it is the
same as it would have been if any of the alternative randomizations associated with the data permutations in the reference set
had been the obtained randomizations. Employment of a permutation group was necessary but not sufficient to provide a closed
reference set. A valid reference set for random data permutation,
on the other hand, does not need to be the same as it would have
been if some alternative randomization associated with the reference set had been the obtained randomization; a reference set is
valid if it meets this requirement: *the probability of getting the
same reference set is the same, no matter which member represents the obtained randomization.* A reference set meeting this
requirement will be called a *stochastically closed reference set.*
The validity of random data permutation procedures was discussed
in Section 3.4, but describing the procedure in terms of randomization sets may be helpful. Before an experiment, an experimenter randomly selects 1,000 of the possible randomizations and
uses the last one as his random assignment procedure. All randomizations in the subset of 1,000 have the same probability of
being the obtained (last picked) randomization, so of all random
samples that provide the particular 1,000 randomizations, it is
equally probable that any of those randomizations will be the
randomization actually employed in the experiment. When the
data are collected for the obtained randomization, that data permutation permits the use of the numerical values associated with
each experimental unit to transform the other 999 randomizations

in the subset into data permutations which, under the null hypothesis, would have been obtained. As they all had the same probability of being the obtained results, the computation of a P-value as the proportion of data permutations with test statistic values greater than or equal to the obtained value is valid.

12.15   CHUNG AND FRASER PERMUTATION GROUP EXAMPLE

In the Chung and Fraser (1958) example of a permutation group, there were several subjects in each of two groups, and the test was a test of the difference between the two groups. (Whether the comparison was of two different kinds of subjects or of two different treatments was not specified.) Suppose there were eight subjects assigned to two groups with four subjects per group. A group of the kind Chung and Fraser proposed concerns moving the first two measurements to the end ("rotation of pairs") until no new data permutations are produced, and may be illustrated by the reference set in Table 6. In the correlation example in Section 12.5, there was a rule for ordering the obtained data, as the drug dosages were ordered. But the ordering of measurements within A and B for the Chung-Fraser example was not defined. The typical application of a randomization test of a difference between treatments does not need to take into consideration the order of the measurements within the treatments, as the within-treatment ordering does not

TABLE 6   A Reference Subset Produced by
Rotating Pairs of Measurements

Treatment A				Treatment B			
X_1	X_2	X_3	X_4	X_5	X_6	X_7	X_8
X_3	X_4	X_5	X_6	X_7	X_8	X_1	X_2
X_5	X_6	X_7	X_8	X_1	X_2	X_3	X_4
X_7	X_8	X_1	X_2	X_3	X_4	X_5	X_6

matter. But the Chung-Fraser group does require a specification of order within treatments.

Let us consider why ordering within A and B is necessary
for the Chung-Fraser group. The reason is that the reference
set will vary according to the initial ordering of data within the
A and B treatments. The divisions of experimental units between
treatments represented in the reference set depend on the initial
ordering. Notice, for example, that if the sequence of measurements for the first data permutation in Table 5 was reversed
within treatment A, then the second data permutation would have
measurements X_1, X_2, X_5 and X_6 under treatment A, a data
permutation that is not included in the subset in Table 5.

A specification of the order of measurements within treatments is necessary, but if a systematic procedure is used, it
does not appear that it could be on the basis of the data as
such. Ordering of the obtained results on the basis of size of
measurements within A and B does not ensure ordering by size
for the data permutations generated by rotation of pairs of measurements, nor, apparently, is there any other systematic ordering procedure, except ordering the measurements according to
the order of assignment of subjects within A and B, that would
provide an unambiguous group of permutations. If one ordered
the measurements within treatments according to the sequence in
which subjects were assigned to the treatment, the group of data
permuting operations then could consist of (1) ordering the obtained measurements according to the order of assignment of subjects providing the measurements, followed by (2) rotation of
pairs to provide the reference set of four data permutations.
Let us consider the randomization-referral approach to determining significance for this situation. We consider not only the $8!/$
$4!4!$ possible divisions of the eight subjects between A and B but
also the $4!$ orders of assignment within each division, which
together provide $8!/4!4! \times (4!)^2 = 8!$ distinctive randomizations.
Application of the "rotation of pairs" group to each of the randomizations divides the set into subsets of four randomizations
each. The knowledge of the ordering of assignment to treatments
identifies the obtained randomization, enabling us to use the subset of randomizations of which that randomization is a member
for generating our reference subset of data permutations.
Whether the randomization-referral procedure or the systematic
data-permuting procedure is used to determine significance, information is required which is not required usually for data-permuting procedures: knowledge of the randomization associated
with the obtained data permutation. Just knowing which four

measurements came from A and which four from B is not enough for the Chung-Fraser group approach.

Now, let's consider an alternative approach that uses only information ordinarily recorded: unordered measurements for A and B. We order measurements within A and B for the obtained results randomly, providing this group of data permuting operations: (1) order the obtained A and B measurements randomly within A and B, followed by (2) rotation of pairs to obtain the reference set of four data permutations. With this approach, the set of randomizations consists of 8!/4!4! divisions of subjects between A and B, the order of the assignment being irrelevant. For the systematic data-permuting approach, discussed earlier, there was a one-to-one correspondence between randomizations and data permutations, but for this alternative approach there are 8! possible data permutations but only 8!/4!4! randomizations associated with them. Each data permutation is associated with a particular randomization in conjunction with a randomly determined ordering. In the first approach, any of the four randomizations (ordered assignments) associated with the four data permutations would, under the null hypothesis, have provided the same reference set of four data permutations. On the other hand, in the second approach, any one of the four randomizations (unordered assignments) associated with the four data permutations would not necessarily generate the same reference set of data permutations; however the probability is the same (namely, $1/(4!)^2$) for each of the four randomizations associated with the data permutations that random ordering of data for the randomization, followed by rotation of pairs, would generate the same reference set of data permutations. Random ordering plus systematic rotation of pairs form a *random group*.

12.16   ALTERNATIVE REFERENCE SETS FOR A COMPLEX DESIGN

Let us now consider the testing of treatment effects when subsets of treatments are randomly selected, as in Section 11.3, for a test of matching or proximity. Consider an experiment in which two treatments are randomly selected from a population of three possible treatments, and the subjects are assigned randomly to the two selected treatments, with the same number of subjects assigned to each treatment. (Equal sample sizes are assumed here so that assignment of sample sizes would not unduly complicate the following examples.) The responses are

TABLE 7 Primary Reference Set of Data Permutations

Treatments		Treatments		Treatments	
A	B	A	C	B	C
p data permutations		p data permutations		p data permutations	

recorded, and the obtained test statistic value is computed. We will consider four types of analysis from the standpoint of permutation group theory.

Type 1 Analysis

Let us first consider the use of the primary reference set of data permutations associated with all possible assignments. That set would consist of 3p data permutations, as shown in Table 7, derived from the obtained results. For example, if the obtained data permutation was: A: 3, 5; C: 4, 9, there would be corresponding data permutations of A: 3, 5; B: 4, 9 and B: 3, 5; C: 4, 9 for the same division of subjects but between different pairs of treatments. In this case, p would be 4!/2!2! = 6. All of the 3p randomizations are equally probable and the permuting of the data provides the alternative results according to the null hypothesis. Thus, this type of analysis provides a primary reference set and thus is valid.

Type 2 Analysis

Next, consider a reference set produced by systematic selection of all treatment pairs that could be drawn, followed by random division of the data for each treatment pair 1,333 times between the two treatments, with the same number of subjects for each treatment. Thus, we have 3,999 randomly generated data permutations plus the obtained data permutation, making a reference subset of 4,000 data permutations. The list of data permutations for the pair of treatments providing the obtained results contains 1,334 data permutations, including the obtained data permutation, whereas each of the other two pairs of treatments have only 1,333 data permutations. This reference set of 4,000 data permutations is not closed. Each of the 1,334 data permutations for the treatment pair associated with the obtained results had the same probability of providing the same reference set of 4,000 data permutations, but it would be impossible to generate that reference set if any of the other 2,666 data permutations had been the obtained data permutations, because there then would have been 1,334 data permutations under a treatment pair presently containing only 1,333 data permutations. Thus, we do not have a systematic or a stochastically closed reference set. On the other hand, we would have had a stochastically closed reference set and thus a valid randomization test if our rule had been to permute the data for the obtained pair of treatments 1,332 times and the data for each of the other two pairs of treatments 1,333 times, providing the same total number of data permutations (including the obtained data permutation) for each treatment pair. Then, it would be equally probable that the same reference set of 3,999 data permutations would be produced, no matter which of the 3,999 was the obtained data permutation. The set therefore is stochastically closed and, consequently, valid.

Type 3 Analysis

Another way to generate a reference set is to randomly select one of the two treatment pairs *not* containing the obtained data permutation and systematically produce p data permutations for that treatment pair and p data permutations for the treatment pair containing the obtained data permutation. The 2p data permutations provide a valid reference set to serve as the basis for a randomization test. Notice that this data-permuting procedure is equivalent to randomly selecting two of the treatment pairs prior to the experiment for providing the reference set and then randomly deciding to which of the two treatment pairs the subjects will be assigned.

Type 4 Analysis

A fourth type of reference set can be produced by a procedure similar to the one for Type 3 analysis. As in Type 3 analysis, we randomly select one of the two treatment pairs *not* containing the obtained data permutation and use data permutations from that treatment pair and from the treatment pair containing the obtained data permutation to provide a reference set. In this analysis, however, the data permutations within the two treatment pairs are randomly, rather than systematically, generated. We randomly produce 499 data permutations for the treatment pair containing the obtained data permutation and 500 data permutations for the other treatment pair, making a reference set of 1,000 data permutations, including the obtained data permutation. This is a stochastically closed reference set and can serve as the basis for a randomization test.

12.17   SUMMARY

Little theory is needed for justifying randomization tests based on primary reference sets, but restricted null hypotheses must be tested by use of reference subsets, and the rationale for such tests requires careful examination. Permutation groups are essential to data-permuting tests, and the randomization-referral procedure is useful for assessing the validity of data-permuting tests. Valid reference sets must be closed, either deterministically or stochastically. Stochastically closed reference sets for random data permutation and for data permutation with both systematic and random components are discussed in Sections 12.14, 12.15, and 12.16. The determination of validity of various tests by consideration of whether their reference sets are closed is discussed throughout the chapter.

REFERENCES

Chung, J. H., and Fraser, D. A. S. (1958). Randomization tests for a multivariate two-sample problem. *Journal of the American Statistical Association 53*, 729–735.

Dwass, M. (1957). Modified randomization tests for nonparametric hypotheses. *Annals of Mathematical Statistics, 28*, 181–187.

Edgington, E. S. (1969). *Statistical Inference: The Distribution-free Approach.* McGraw-Hill, New York.

Edgington, E. S. (1983). The role of permutation groups in randomization tests. *Journal of Educational Statistics, 8*, 121–135.

Kempthorne, O., and Doerfter, T. E. (1969). The behaviour of some significance tests under experimental randomization. *Biometrika*, 231–248.

13
General Guidelines

This final chapter will provide some general guidelines for making effective use of randomization tests.

13.1 MAXIMIZING POWER

The versatility of the randomization test procedure provides the experimenter with so many valid alternatives in conducting experiments and analyzing the data that it is important for him or her to have a basis for choosing procedures that are most sensitive to treatment effects.

Homogeneity of Subjects

The experimenter should select subjects that are as homogeneous as possible to make sure that between-subject variability does not mask treatment effects. Homogeneous groups of other kinds can be used to test the generality of a finding for a homogeneous group used in an earlier experiment.

Statistical Control over Between-subject Variability

Direct control over between-subject variability should be exercised through selection of homogeneous subjects, but when the subjects must be heterogeneous statistical control may be necessary. Treatments-by-subjects designs can be used to control for heterogeneity of subjects, as can analysis of covariance and partial correlation. Repeated-measures designs and single-subject experiments also are useful for minimizing the unwanted effects of between-subject variability, thereby providing a more sensitive experiment.

Number of Subjects

As the number of subjects increases, the chance of detecting treatment effects also increases. In single-subject experiments the chance of detecting treatment effects increases as the number of replications of treatment administrations increases.

Unequal Sample Sizes

Subjects should not be left out of an experiment simply to ensure equal sample sizes, equal cell sizes, or proportionality of cell frequencies. However, when there is a disproportionality of cell frequencies for a factorial design, the test statistic for nondirectional tests should be the special one given in Section 6.12, or else the data should be transformed according to the procedure in Section 6.14.

Restricted-alternatives Random Assignment

Sometimes more subjects are available when the assignment possibilities are allowed to vary from subject to subject. When restricted-alternatives random assignment is employed, however, the procedure in Section 6.12 or in Section 6.14 should be used for nondirectional tests.

Equal Sample Sizes

Although samples should not be restricted in size to ensure equality, equal sample sizes are preferable. With equal sample sizes there are more potential assignments and, consequently, a smaller P-value is possible. For example, for independent-groups assignment, there are $10!/5!5! = 252$ possible assignments of five subjects to one treatment and five to the other, whereas there are only $10!/3!7! = 120$ possible assignments of three subjects to one treatment and seven to the other. Another advantage of equal sample sizes is that statistics and experimental design books frequently give computational formulas for equal sample sizes only.

Counterbalancing

Counterbalanced designs, such as those discussed in Sections 5.13 and 10.16 put constraints on the random assignment possibilities and thereby limit the smallness of the P-values that can be attained. Such designs should be avoided unless there is no way to eliminate a strong sequential effect that is expected. Sometimes increasing the time between successive treatments will minimize such effects, serving the function of counterbalancing without restricting the random assignment possibilities.

Treatments-by-subjects Designs

Treatments-by-subjects designs, as discussed in Section 6.8, can provide effective control over between-subject variability, but there are fewer possible assignments than with one-way ANOVA. Consequently, unless there is interest in the treatment effect on each type of subject separately, the experimenter should strive for homogeneity of subjects to make this design unnecessary.

Precision of Measurement

Measurements should be as precise as possible. There is no necessity for reducing precise measurements to ranks to perform a rank-order test; the same kind of test can be performed with raw data when significance is determined by data permutation. Any degrading of precision of measurements should be avoided.

Precision of Test Statistics

To make effective use of the precision of the measurements, the test statistic should take into account the precise measurements, not just the rank order. Means and standard deviations are more precise measures of central tendency and variability and, consequently, they are more appropriate test statistics or test statistic components than are medians and ranges.

Composite Measurements

In a multivariate experiment when measurements are combined to form a single composite measurement, the procedure should be such that the significance value is independent of the units of measurement. One means of accomplishing this is through the use of z scores, as in Section 7.4.

Combining Test Statistics over Dependent Variables

When test statistics are computed separately for each dependent variable in a multivariate experiment and they are added together to get an overall test statistic, the test statistics that are added should be such that the P-value is independent of the units of measurement. For example, in Section 7.5, F is such a test statistic but $\Sigma(T^2/n)$ and SS_B are not.

Directional Test Statistics

For categorical as well as continuous measurements, one-tailed tests should be used when there is a good basis for a directional prediction. A directional test statistic, such as the total frequency in the lower right cell, may be substituted for the nondirectional chi-square test statistic with a 2 × 2 contingency chi-square test. There is no factorial t counterpart to factorial F for main effects, but when there are only two levels of a factor, a one-tailed test for the main effect of that factor over all levels of the other factors can be carried out by using T_L, the total of the measurements for the level predicted to provide the larger measurements, as the test statistic. One-tailed tests are preferable to two-tailed tests when the direction of difference between means or the direction of relationship, as in correlation, can be predicted. Occasionally, however, journals require that two-tailed P-values be reported. When there are more than two levels of a treatment and a directional prediction is made, trend tests like those in Chapter 9 can be used. When trend tests are used they should be designed to accommodate the full specificity of the prediction.

Control over Systematic Trends

When a subject takes a number of treatments in succession, in either a single-subject or a repeated-measures experiment, and there is a general upward or downward trend over time, a test that compensates for such a trend will be more powerful than one that does not. For instance, one can use a randomization test based on difference scores, analysis of covariance with time as a covariate, or some type of time-series analysis.

Random Sampling of Levels of Treatments

In Section 11.3 it was shown that random selection of treatment levels in conjunction with random assignment can substantially increase the power of a test. There are only certain situations where this is possible, but careful thought should be given to this possibility when there are only a few experimental units.

Combining of Probabilities

The P-values for pilot studies should be combined with those for the main study to increase the power of investigations. Similarly, the P-values for different experiments testing the same general H_0 should be combined. P-values also can be combined over replications of single-subject experiments with different subjects.

13.2 COMPUTER PROGRAMS

Modifying Programs

The permuting procedures in the programs given in this book correspond to the common types of random assignment and, as a consequence, the programs can be readily modified to serve in many experiments. For instance, a program can be changed by altering the test statistic to make it sensitive to differences in variability instead of differences in size of measurements. Parts of one program can be combined with parts of another to make a new program, as was done in deriving the correlation trend program, Program 9.2. Program 9.2 is appropriate for the assignment associated with one-way ANOVA, so the data-permuting component is that of Program 4.3. The test statistics, however, are the same as for the correlation program, Program 8.2, so the component of Program 8.2 which concerns the computation of test statistics was incorporated into Program 9.2. Programs that are not readily derivable from ones in this book are ones for special random assignment procedures, like restricted-alternatives random assignment, random assignment for counterbalanced designs, and some of the assignments for single-subject operant research designs described in Chapter 10.

Equivalent Test Statistics

For both systematic and random data permutation, simple test statistics that are equivalent to the more conventional test statistics are useful. Not only do they reduce the computer time but they also make it easier to write and check programs.

Random Data Permutation

Random data permutation programs are more practical than those with systematic data permutation, because little computer time is required for moderate or large samples. They are perfectly valid in their own right; the closeness of the P-value given by random data permutation to that of systematic data permutation is related to power alone, not to validity. Although the power is less than the power of systematic data permutation, the discrepancy is small with even a few thousand random permutations and can be reduced even further by increasing the number of random permutations employed. (See Sections 3.4 and 3.9.) When a new type of random assignment is used, it is easier to develop a random data-permuting procedure that will match the assignment procedure than to develop an appropriate systematic permuting procedure.

Systematic Listing Procedure

For systematic data permutation programs, a systematic listing procedure is required. For conventional types of random assignment, the data permutation procedures used in the systematic data permutation programs in this book are the required listing procedures. However, for unconventional types of random assignment, a new systematic listing procedure must be devised first, and then incorporated into the program.

Elimination of Redundant Data Permutations

For systematic data permutation programs, the reduction in the required number of data permutations can be considerable for certain equal sample size designs when redundant data permutations are omitted. When sample sizes are equal, the experimenter using a systematic data permutation program should check to see if the symmetry of the sampling distribution of data permutations allows him to alter the systematic listing procedure so that only a fraction of the total number of data permutations have to be listed. This, however, is a refinement which, like the formation of equivalent test statistics, is not essential to a systematic data permutation program and should be carried out only if a program is to be used repeatedly.

Output

A computer program for a randomization test should provide more than a P-value. It should also provide the value of t, F, or other conventional test statistics for the obtained results when the program is for determining significance of a conventional test statistic by data permutation. Equivalent, simpler test statistics may be used to determine significance, but the ordinary test statistic should be determined for the obtained results so that it may be reported in the research report. Additionally, other statistics for the obtained results, such as means and standard deviations, may be appropriate to compute, sometimes, for reporting the results of the analysis.

Testing Programs

Sections 3.8 and 3.9 discuss the testing of programs. Test data for which the P-value is known, like the data accompanying the programs in this book, should be used. The systematic programs should give the exact P-value and the random programs should provide a close approximation, given 10,000 random data permutations (see Table 2 in Chapter 3). It is best to test the programs with data that do not give the smallest possible P-value because there are several ways for a program to be incorrect and yet give the smallest P-value correctly. In testing a program it is frequently helpful to have all or some of the data permutations and test statistic values printed out in addition to the P-value for the test data.

Index